Guido van der Groen

ON THE TRAIL OF EBOLA

My Life as a Virus Hunter

 LANNOO

ABOUT THIS BOOK

On the Trail of Ebola follows a timeline that includes the most important events concerning the Ebola virus, but also a few milestones in the history of the Congo and the development of virology and science in general. The tapestry of virus hunter Guido van der Groen's life is woven between the facts. The story of Ebola (and of AIDS) cannot be told without some technical explanations. Names, concepts, and abbreviations that demand clarification have an asterisk (*) and are defined in a glossary at the end of the book.

The author mapped out the timeline on the basis of scientific and anecdotal literature, along with the tsunami of reportage on ebola since the summer of 2014. The chronology is based on sources researched and news collected by the author and was completed as late as possible before the book went to press. The sources are stated in a list, also at the back of the book.

On the Trail of Ebola contains no new scientific findings. But the author did everything possible to uncover and check the relevant facts—and to do right by the scientists involved, wherever they may be. At the same time the author has used language that can comfortably be read by non-scientists. Naturally, some compromises had to be made. The wide awake reader can notify the publisher of final inaccuracies or flaws.

Much historic photographic and video material connected with Ebola was specially made available to the readers of this book by The Institute for Tropical Medicine. Among this material are images that Guido van der Groen himself captured in Zaïre during the first outbreak in 1976. We heartily recommend that the reader takes a peek at www.onthetrailofebola.com during or after reading this book.

CHAPTER 1

MONUMENT TO A MISTRESS

The greatest Ebola outbreak in history becomes world news,
and I experience my busiest period as an experienced virus
hunter. I delve into my archives in order to reconstruct
a diary that relates how Ebola determined the courses
of my own life and those of my nearest and dearest.

Tuesday, August 12, 2014

A Cetti's warbler settles on the shrub next to the little pond in my garden and ruffles his red-brown feathers, unaware of the observer behind the sash window. I don't need to be quiet because with the uninterrupted traffic on the E19 a couple hundred meters away, even a herd of trumpeting elephants wouldn't alert the warbler. Yet I feel trapped while I stand staring at him from such an exposed position. I don't see him that often in my Kontich oasis, this beautiful songbird that is named after the Italian Jesuit, zoologist, and scientist Francesco Cetti.

It promises to be a simmering hot day and I must let off some steam. My 71-year-old body and brain tire more quickly than twenty years ago, I must admit. Over the last few weeks the cell phone of this "professor in retreat" has worked overtime and I have often slept badly. Today, at an impossible hour, I wrote in my jotter: "12 August toll epidemic* over 1000 deaths: 1031 deceased out of 1848 infections.*" Since April I have been following the most recent figures on the website of the World Health Organization (WHO) almost every day. Paging back through my jotter I note that I first remarked on this outbreak* on April 9: "The virus* was diagnosed in Conakry, the capital of Guinea, with 157 cases clinically suspected, of which 101 were fatal. In Liberia there are 10 deaths. Also in Sierra Leone and Mali the first infections are a fact." Apparently at that time I wasn't seriously upset... But that was April, and the only striking things then were that the virus had, for the first time, appeared in West Africa and not in Central Africa, and had hit Conakry, a large town. All previous outbreaks up to the last one at Kikwit in 1995 had begun in the countryside.

The Cetti's warbler now has me in his sights and spreads his wings. I have also often done that in my career, and usually the cause was a virus.

Fall 1976

As a young man in my thirties at the Antwerp Institute for Tropical Medicine (ITM),* I prepare blood samples for microscopic investigation. Three years earlier I had graduated as a doctor in biochemical sciences and joined battle as the assistant of Professor Pattyn in the ITM. Pattyn made me study the bacilli that cause leprosy and tuberculosis.* But these began to blunt my wits, and one day, when my boss was close by, I sighed: "Give me a few viruses." And so it was on a morning in September 1976 that I, together with a few other researchers, opened a cheap thermos flask and came face to face with blood samples from a Belgian mission sister in Zaïre. Her name was Sister Myriam, and she already had a high temperature and was hemorrhaging when the blood was drawn. At the same time the vials reached Antwerp, the sister passed away. Two other Belgian nuns and tens of Zaireans had already preceded her.

A good week later, when I studied the suspected cause under the electron microscope, I found myself looking at a whopper of a virus. It makes Stefaan Pattyn think of Marburg—the giant unknown virus that popped up in 1967 to make the scientific world shiver with angst. But this, it will appear, is something else.

Not even two weeks later I sit in an aircraft flying to Zaïre, where I, together with the other members of an international research team, will hunt the virus. During this period—the most formative of my life—I start an extramarital affair. Ebola is the name of my cold lover, and as it so often goes with mistresses, she cloaks herself in mystery. For years I will doggedly seek the animal host in which the virus shelters between outbreaks until its next opportunity to ambush people and great apes—and cut them down within ten days.

August 2014

Never will Ebola hit so hard as during the epidemic that begins on December 6, 2013, with the death of a two-year-old toddler in a Guinea village. By today, August 12, 2014, Mademoiselle Ebola has claimed more than one thousand lives. Only on March 25, 2014, did the WHO announce that there was an Ebola outbreak in the region. The American research institution Centers for Disease Control and Prevention (CDC*) also went public that day, with a report about 86 cases and 59 deaths in Guinea, supposedly as a result of the Zaïre Ebola virus. At that moment experts assumed that they were dealing with a classic Ebola pattern, in which the number of infections first rises spectacularly, only to plummet just as quickly somewhat later, when the heart of the fire is extinguished.

Two weeks after the first publications experts once again find their hearts pounding with anxiety, and I resume jotting down figures. The internet reports follow each other ever quicker and the statistics rocket. West Africa—that it should be happening there intrigues me.

Because I do not trust blindly in the memory of a 70-plusser, I hunt out the 2011 thesis of the Frenchman Paul Roddy, whose studies at the ITM involved listing and describing all Ebola outbreaks since 1976 with the number of confirmed infections and deaths. I sat in the jury when Paul defended his thesis, and I recall that his information was trustworthy. My finger runs over the list. Nowhere is a West African country. With a pencil I circle the highest numbers: Yambuku (Zaïre, 1976): 318 cases and 280 deaths; Kikwit (Zaïre, 1995): 315 cases and 254 deaths; Gulu (Uganda, 2000): 425 cases and "only" 224 deaths.

True enough, in June I do not yet see any Cetti's warblers in the garden, but the first alarm calls nonetheless appear in my jotter. On July 4 the WHO mentions 470 deaths and the West African outbreak is now the deadliest ever. The Belgian TV channel Canvas has found my telephone number and asks if I will come and explain events on the program Terzake, a show consisting of reports, debates, and conversations. It becomes the first interview of a long series that I will give in this outbreak, for both domestic and foreign media. It is also the stimulus to draw a deep breath and dive into the archive of my thirty-year academic career. In three decades a person gathers quite a heap! My career in pursuit of Ebola has taken me to all continents except Antarctica.

Through all these years I wrote my travel impressions on pieces of glossy cardboard that I had clipped to the inside of my breast pockets. After computers became naturalized citizens of the ITM, my secretary, Ciska Maeckelbergh, neatly retyped all the information from the cards. In 2003, the year I became a pensioner, I switched over to Aurora-notes, which I crammed minutely with black ink. I use a computer to communicate or find things, but I prefer to write my thoughts and commentaries by hand. I note everything that interests me as it happens—things I learn from the media, things in scientific journals, memorable quotes from gripping talks. Such obsessive behavior through all these years seems somewhat autistic, but at the same time I reckon with the possibility that all that trouble will one day be of use. And so it is that my chronicles clearly relate how the previous outbreaks differ from the current one. This is the first time that we see the virus explode with such force. Never before has this happened in three countries simultaneously, and never before in West Africa—not that health care there is any better than in "traditional" Ebola lands such as the Congo. Moreover, Liberia,

Guinea, and Sierra Leone are always licking their wounds after civil wars and military uprisings. They can overlook an epidemic of an unknown disease as a mere toothache.

My list of media appearances lengthens by the day as more events occur related to the virus. On August 12 I give an interview to *The Voice of America*. Two foreign aid workers have been repatriated from Liberia to the United States, and that is big news there. Ever more airlines limit or scrub their flights to West Africa. Here and there in fearful white countries passengers from suspect areas may no longer enter or are subject to strict checks. Opinion makers of all sorts, some more objective than others, intervene in the debate. All the media outlets are full of Ebola. Images of health workers clad in yellow spacesuits invade households worldwide. Questions from journalists rain into my cell phone. Most journalists are not familiar with the virus. Some recall the Kikwit '95 outbreak, but Yambuku '76 means nothing to them. Many interviewers weren't even born then. Only now does Ebola penetrate the collective conscience and evoke many questions. What sort of virus is this? Can you recover from it? What if it also strikes in Belgium?

Over the years I have often thought aloud that at some time I must write a book about Ebola and how this virus determined my career. Now that idea seems firmer by the day. It could give answers to everyone who, with open mouth and knitted eyebrows, finds himself washed away by the deluge of Ebola news. Of course such a book is a titanic undertaking. Perhaps I should seek help.

Dina, my wife, thinks such a book is a good idea. Alert as always, she draws my attention to the sound recordings that I had sent from Yambuku in 1976. They must be lying around somewhere and can probably refresh my memory. In fact even the suggestion does that at once because I remember that during a later trip to the pygmies, I had filled whole cassettes. I also find the diary of my French friend and colleague Pierre Sureau of the Institut Pasteur in Paris. He was there alongside me in 1976 and could call himself the co-discoverer of the Ebola virus.

I begin preparations for my book. Ilse, my oldest daughter, recommends a female friend with a fluent pen. I reread the novel *Sister Veronica* by William T. Close, Bill to his friends. Bill was the American doctor of President Mobutu and the father of actress Glenn Close. He, too, had been in the team of 1976. I sort newspaper cuttings and typed breast pocket cards into piles. Slowly but surely the jigsaw pieces of my memory fall into place.

I come across a poem. As a young buck, on evenings when not a morsel of science was being imbibed but only good whiskey and beautiful music, I actually wrote poetry. There is also a date on the poem. Dates assign time to volumes. But I also like numbers and summaries. I readily put things into systems and procedures, in detail. This characteristic has helped me in the study of Ebola, poxes (mostly smallpox*), hantaviruses,* and HIV,* and it contributed to the fact that at a certain time I was able to build the smallest maximum security laboratory* in the world. In 1983 I had to demolish that lab because the ITM needed space for our new focus: AIDS research. For an Ebola fanatic like me it wasn't easy to start again from the beginning, but that, too, is science: killing your darlings and dedicating yourself to new developments and alliances. With that motto before my eyes, as the chief virologist of the ITM I would build the most expansive network possible, in both scientific and pharmaceutical circles.

The Institute for Tropical Medicine was my window on the world. By studying outbreaks in so many different locations, I had learned to recognize viruses not only as a cause of disease, but also in their other guises. The virus as panicmonger, for example. Or the virus as a cunning character that prevents people learning from history and devising permanent solutions. Perhaps that is the only point of the devastating, unprecedented epidemic in West Africa: that this time Ebola has attracted the attention of the world and of a new generation and held it for months. If politicians act more decisively in the future and are more willing to invest in enduring health care, anywhere on Earth, then my mistress may get a monument. That would be a fine ending, for myself too. Then I can close my archive with my mind at rest.

But let me first tell how it all began.

CHAPTER 2

1942 – 1972

HORSESHOES, AUNT TAANS, AND THE ALMIGHTY WARRIOR

My youth in the shadow of the chapel in Kontich-Kazerne,
V2 missiles, conscience, monkeys in Germany, and riots
in the Congo. And how the siren-call of far continents,
new viruses, and a half-Greek woman shape
my life definitively.

Friday, December 11, 1942

In the midst of World War II Maria Vingerhoets rushes to the maternity clinic of Kontich to give birth to her first and only child.

A day or two later, Willy van der Groen registers a son under the name Guido, patron saint of farmers, coachmen, and vergers. Father is a confirmed atheist and will have nothing to do with saints, but Mother is a practicing Catholic—no easy union in an era when the pastor regularly drops in on his parishioners. In our case he habitually asks the question: "Well then, Maria, when will I see your Willy spend time in my church?"

1942-1947

In the year of my birth, the German army experiments with the supersonic V2 rocket that you can neither hear nor see before it arrives. In late 1944 Germany uses these V2s to strike Antwerp and its environs from the stratosphere. Such a rocket falls not far from my parents' house in Kontich-Kazerne. Meanwhile, I take my first baby steps, gripping the hand of my grandfather. Henri van der Groen—with a small "v," a detail the family feels to be important—is the first Dutchman who settled in Kontich at the beginning of the twentieth century. He is a commercial gardener and creates new plant hybrids. Henri is also the great uncle of Dora van der Groen, who will be one of Flanders's best and most renowned actresses. Apart from being a florist, Dora's father is also a cellist, painter, and humorist. His cousin—my father, Willy—also draws heaps of cartoons during his life, a family tradition that I will maintain with pleasure. One can say that art and humor run more in the family blood than science.

On winter evenings I lie next to my father on the narrow sofa in our living room. His ribcage pushes against my cheek in the relaxed rhythm of his breathing. The only lights are the glowing coals in the grate and the large radio. The only sounds: Father's heartbeat and Jan-Albert Goris, alias Marnix Gijsen, *the Voice from America*. He had been living in the US for years and recently started to read a radio commentary on his life over there, every Saturday evening. The trans-Atlantic connection makes me dream about far continents.

During the last two summers of the war a strange disease appears among Soviet soldiers in the Crimean peninsula of the Black Sea. The soldiers were there helping the local population to gather the harvest, since the Turkish Crim-Tartars disappeared following ethnic cleansing. The soldiers slept in hay on the fields, under the stars. More than two hundred contracted hemorrhagic fever,* a fever combined with bleeding. Later it will be apparent that the cause of Crimean Hemorrhagic Fever* is a virus transmitted by ticks. The Crimean summer fields always swarm with ticks.

Wednesday, April 7, 1948
The World Health Organization* (WHO) is established as the first specialized division of the United Nations (UN). The most important goal of the WHO is to promote the health of all people on earth. From now on, April 7 will always be World Health Day.

1948–1952
I begin my first year in the primary Catholic school of Kontich-Kazerne, at the spot where the library will later stand. I learn to write between lines using a pen that I dunk in an inkpot. The smell and feel of writing with a fountain pen are etched in my long-term memory. Lifelong I will like drying ink and comfortable gliding pens, will fill countless cards and notes, and prefer handwriting to a keyboard.

A year later I travel by rail to Antwerp-Central. My parents, with their diverging religious views, have reached a compromise: to accommodate Father, Mother allows her beloved son to attend a government school; but out of respect for Mother's piety, he will take Catholic religious classes there and not ethics. The religious instruction does not miss the mark. So it is that I become acquainted with the miracle of confession. I learn that I must feel so guilty that I will have the urge to report my sins to someone who will confirm that I have indeed gone astray, and who will consequently make me rattle off a few short prayers as punishment, after which I can once again frolic through life. That is genius.

I confess regularly in the church at Kontich-Kazerne. Right next door is the blacksmith, where I linger and look around. On one occasion I see how the blacksmith pulls up one of Jules's hooves, lays it across his thigh, and fits a brand new horseshoe. I smell burnt horn. The tensed haunches of the farm horse do not flinch. The blacksmith spots me and with a wink he flings a discarded horseshoe on the cobblestones in front of my feet. I scoop it up and walk into the church. There is someone ahead of me. The horseshoe is covered in scratches and blemishes. Those will be my mortal sins! When I am sitting on the confessional stool I will describe all spots and scratches, and during my penitence with every Hail Mary and Our Father a blemish will disappear, so that afterward my soul will shine as brilliantly as Jules's new horseshoes. Following confession the church is lighter than when I walked in, and when I leave I screw up my eyes tightly against the harsh sunlight.

Now and then the parents of my second cousin Dora visit us. Hortensia, "Aunt Taans" to the family, is a jewel of a woman who tells wonderful stories. Such as that about her brother Hugo, the colonist, who has a huge farm "in the Congo," where the black people hunt wild animals with bows and poisoned arrows. She tells us that you must have incredibly strong arms to pull back the African bows. I fashion my own bow and paint concentric circles on an old bedsheet that I stretch across a laundry basket, placing the target at one end of the garden. Thus I was a hunter before I had viruses in my sights.

A war has been raging since 1950 in Korea. The trenches are infested by rats infected with an unknown virus, which will be called "hanta." The rodents shed the virus continuously in their urine and droppings. Soldiers breathe in the dust and become infected. In this way 3,000-odd UN soldiers contract hemorrhagic fever and kidney problems. Ten percent of them die as a result, Belgian soldiers among them.

Saturday April 25, 1953

In the elite British scientific journal *Nature* an article appears that becomes world famous: "A structure for deoxyribose nucleic acid." In 1952 it had already been shown that DNA* is the carrier of inherited characteristics, but now it is established that a DNA molecule has the structure of a double helix, two long strings that are bound together by so-called base pairs, like a ladder that you have twisted to the left at its top end, and to the right at the bottom. Usually only the names of Francis Crick and James Watson are connected with this discovery, but as so often happens in science, the true story is more complex. This will also be the case after the discovery of the Ebola virus and HIV.

1954–1955

Because my parents think that everyone should set the bar as high as possible, I move to the Atheneum in Antwerp, where I follow the Latin-science curriculum. It soon transpires that I am just an average scholar, but in one subject I am outstanding: religion. Mother is delighted. One of my fellow students is Eddy Van Vliet, who will become well known as a poet and lawyer. Eddy follows the Greek-Latin curriculum, but we share some subjects. Due to his exceptional feeling for language he is the darling of our Dutch teacher. That puts my nose out of joint. At this time poetry is nothing to me. In 2002 Eddy Van Vliet will die much too young.

In the Athenaeum another literary celebrity crosses my path: Aster Berkhof, pseudonym of Lode Van Den Bergh, and also our Dutch teacher. We envy him, not because he is a productive author—at the time of writing Berkhof is 94 and has written 105 books over 70 years—but because our teacher shares his life with TV-announcer Nora Steyaert! One morning he enters the class yawning. During my talk about the conquest of Mount Everest by Sherpa Tenzing Norgay and Sir Edmund Hillary in 1953, Mr. Van Den Bergh falls asleep.

1956

In the Belgian Congo, Marcel Dewispelaere, a doctor from Bassevelde, and Roger Audenaert, a medical assistant from Zelzate, observe that a deadly disease has broken out in the village of Bili, in the Bondo district of the Eastern Province. Marcel and Roger travel to the village. Upon arriving they encounter tens of people who are still sick, and the mortal remains of about thirty victims. Most of the bodies display bloody wounds and are bleeding from the mouth and anus. The Flemish health workers first think of smallpox, but they drop the theory when they find the scratch marks of vaccinations.* They question patients and family members.

In a letter about their findings Roger Audenaert writes the following: "The sickness began with a temperature, mostly followed after ten days by bleeding from the nose and mouth and sometimes also bloody diarrhoea. The patients couldn't swallow their food properly. People who lost blood died within ten days. Others had no hemorrhagic symptoms. They began to eat after ten to twelve days and were cured after suffering three weeks of sickness. At least 215 people died, among which were two nursing assistants. Three quarters of the dead were men and most of the victims were adults. The youngest victims were 12 years old. All patients belonged to ten families which lived in Bili or within an area up to 20 kilometers above the Bili River. The zone was immediately placed in quarantine. Roadblocks were set up and no one in the region around Bili was allowed in or out. About 80 patients were isolated in a cotton shed. After 12 days almost all of them had died. Graves were dug in the vicinity of the shed. The dead were buried in the blankets upon which they had lain. The helpers who touched patients or bodies wore rubber gloves. The burial place was guarded by soldiers but some families bribed soldiers and took their dead family member back home at night, to rebury him or her according to the traditional rites. The outbreak lasted roughly five weeks."

Audenaert had also speculated about the cause, which he thought possibly came from the area or surrounding wilderness. "The village lay in the transition zone between tropical forest and Savannah. There was a coffee plantation in the neighborhood. Great colonies of bats lived in this area, but according to rumor the local population did not eat these animals. By contrast the people were crazy about monkeys."

Roger Audenaert sent this letter to my colleagues Bob Colebunders and Jef Van Den Ende of the ITM in May 1995, probably after he had heard of the Ebola outbreak in Kikwit, which had surfaced not long before. Colebunders and Van Den Ende concluded—and I agree—that the events of 1956 that Mr Audenaert describes put one very much in mind of the outbreak of a filovirus* twenty years before Yambuku! Moreover, Bili lies 360 kilometers to the north-east of Yambuku, thus not so very far away. In April 2015 *The Lancet* would publish a short article that my colleagues wrote based on the letter sent to them by Roger Audenaert in 1995.

Back in 1956 a youngster of 13 presents himself to a Belgian colonial family doctor in Stanley Town. The child shies away from light and complains of fever, headache, vomiting, and painful joints. Shortly thereafter the general practitioner also falls ill, as well as two colleagues in a laboratory in the Ugandan town of Entebbe, where blood samples from the young lad and the doctor have been sent for investigation. The virus is baptised Congo virus. In 1967 it will transpire that this limited outbreak in the Congo and that which afflicted the Soviet soldiers in the Crimea were caused by the same virus. For many years Western and Soviet scientists discussed the name of the virus. Eventually they settled on Crimean-Congo Hemorrhagic Fever (CCHF),* in the sequence of the outbreaks: first the Crimea, then the Congo. In the Cold War it is one-nil to the Soviet Union, which had turned this into a political question. Important discoveries have many mothers... The second half of the 1950s is indeed a fruitful period for virology, with the discovery of yet more arboviruses (viruses that are transmitted by arthropods, including mosquitoes), such as chikungunya virus (1955), the sindbis virus (1956), and the o'nyong-nyong virus (1959).

Thursday, June 30, 1960

Independence! The Belgian Congo becomes independent. In his speech during the power transfer ceremony the Belgian King Boudewijn praises Leopold II and the colonial administration. Then Prime Minister Patrice Lumumba reacts with a sharp speech against the Belgian government. The colonists do not know what will happen. Many leave. For health care the results will not be helpful.

July–October 1960

I am finished with high school. In half a year I will be eighteen and it is time to decide what I will do with my life. I want to be a pilot, like so many youngsters, but in my case this is more than a childhood cliché. I am fascinated by flight and aircraft. Not so long ago I convinced my little friends to take their seats in my imaginary aircraft and I flew them around the world, or I kept myself busy all day assembling an Airfix model "Spitfire." I seek information here and there about pilots' training, but the heavy mathematics involved scares me off. To become a priest–not an extraordinary idea in those times–is in my case even less of an option. On my sixteenth birthday–in spite of horseshoes and my mother's devotion–I lose my faith definitively. It is a pity that none of my friends or those in my circle of acquaintances were going to study medicine, because that was another calling that could have seduced me.

My parents give me complete freedom to find my own route and do not suggest that I follow in my father's footsteps. He is a mentor and a comrade, but his career does not beguile me. He studied commerce and became the head of the administration of the Antwerp Gas Company. Willy van der Groen was definitely one of the first Belgians who could work with computers. At the end of 1950 the gas company decided to automate the billing of gas consumption. At Expo 58 in Brussels we could wonder at such a computer, a monster that would fill two living rooms and swallowed punch cards. Gevaert was the first business in the area to equip itself with such a beast, an IBM machine. During weekends my father went to Gevaert to learn how to program it and put information on punch cards. At home he continued to practice, a time-robbing activity that demanded his utmost concentration. I was just a nuisance when he was busy with it, but I was fascinated by the patience and precision with which father worked. But even something to do with computers didn't suggest itself to me. The fields of Informatics and Computer Science didn't yet exist. And moreover, I would have had to study mathematics, and with that I could just as well have become a pilot.

Eventually I chose chemistry. That decision was connected to the chemical companies that are sited in and around Antwerp, and the prospect of job security. I register at the national university of Ghent (Rijksuniversiteit van Gent). At that time you couldn't study chemistry in Antwerp. You could do so in Brussels, but the enrollment fee there was much higher. Leuven was too far for me, and surely too Catholic. In my freshman year at Ghent we start with one hundred and twenty students. I quickly note that ten of them are girls.

Monday, April 17, 1961

The medical faculty members of the Belgian universities establish the Medical Tropical Foundation, Fométro.* Thanks to Fométro, many medical activities in the Congo that fell prey to political feuds and sabre rattling immediately after independence remain going concerns. For almost eighty years Belgium—first Leopold II and thereafter the Belgian state—had almost every imaginable colonial crime on its conscience, exploiting the riches of the Congolese earth, educating the population but being sure to keep them powerless. Granted, at the moment of independence the Congo had the best public health service in Africa. That system is now on the slippery slope. Belgian doctors and researchers return to their homeland and wish to work in the Institute for Tropical Medicine in Antwerp. ITM welcomes the bacteria, virus, and parasite hunters of the "Out of Africa group," with their backpacks full of tropical experience. Other colonial health workers move to Janssen Pharmaceutica or Belgian universities. For the Congo, such an exodus of so much medical expertise is a bloodletting, but happily not all doctors and scientists desert the burning ship, and from now on they can count on Fométro to keep things going.

June 1964–June 1965

After four years of study, just 34 of the original 120 students remain. We are very proud of our degrees, Dina and I. In a country where women have only enjoyed full voting rights for twelve years, it is not taken for granted that a young woman will study at university, let alone that she will choose such a male bastion as science and chemistry. So she is someone to be reckoned with, and that is the least one can say about Constantia "Dina" Pitridis. Dina, the product of an Antwerp mother and a Greek father, is a clever woman who finished with better marks than I did. Although, I admit, it was her beautiful legs that set my head spinning.

During our undergrad years we have become a couple and we are both given the opportunity to begin doctoral studies in biochemistry with Professor Massart. Dina applies herself to plant hormones, and I concentrate on the study of enzymes. As the assistant of Professor De Bruyne I work in the laboratory for general and biological chemistry. I also organize practical exercises in inorganic chemistry for first-year medical students. Teaching and supervising students suits me well. I think no more about applying for a job at a chemical company.

The enzymes that I wish to study for my doctorate must be isolated from bacteria. The renowned Jeff baron Schell, molecular biologist and for many years the director of the Max Planck Institute, teaches me how I must set about this and how to culture bacteria. He died in 2003.

Thursday, November 25, 1965

In the Congo, General Mobutu seizes power after a first coup d'état.

Saturday, October 1, 1966

Dina and I marry.

August–September 1967

At the beginning of August three co-workers of the vaccine company Behringwerke fall ill in the German town of Marburg, in the region of Hessen. They have muscle complaints and headache. Although it is summer, the first thought is early flu, but the symptoms rapidly worsen: bloodshot eyes, vomiting, pain upon touch, rising temperatures. Ten days after the first signs of illness, the three are lying in the hospital. Due to their extreme throat pain they must be fed by intravenous drip. They also have acute diarrhea and are bleeding from body openings and places where injections are given.

Meanwhile it becomes known that other people have similar symptoms. In all, there are 31 cases: 23 in Marburg, 6 in Frankfurt, (90 kilometers to the south), and 2 in Belgrade (in the former Yugoslavia). Of the 31 people, 25 are involved in the production of a live polio* vaccine, for which the tissues of infected monkeys from Africa are utilized. The 24 Germans and the Yugoslav veterinarian have all killed monkeys or dissected monkey cadavers. The 6 other patients were infected secondarily in the university hospitals of Frankfurt and Marburg, among other places, through needle punctures or sexual contact. There is also a mortuary technician who accidentally cut his forearm with a knife. Eventually 7 people will die, all primary infections. The brutality of this unknown infection alarms and upsets the scientific world. I am not yet a member of this community, but Stefaan Pattyn, my future boss, certainly is.

November–December 1967
The evildoer is identified and gets the name "Marburg* virus," after the town where most victims worked. It becomes clear that the virus entered Europe via Ugandan green monkeys, which are used by the Behringwerke to develop polio vaccines. It is a new virus that will only be officially classified as a member of the filoviridae (threadlike viruses) in 1982.

1968–1969
In the 1960s military service is still compulsory for every young man in Belgium. I am granted postponement for my studies but not cancellation, and in 1968 I must do my service. I undertake my duty in the radiology division of the Ghent military hospital as "Candidate Reserve Officer–Technician." For months I take X-rays with equipment that has not been properly maintained for years. I do not want to know how much radiation I clocked up there, but it is a miracle that I beget two bright daughters after this.

In January 1969 Laura Wine, an American mission nurse in the Nigerian hamlet of Lassa, suffers hemorrhagic fever and dies. A week later Charlotte Shaw, the colleague who nursed Laura, also dies from the same illness with an unknown cause. The events in Lassa make world news, yet the characteristics of the disease are not new. Indeed, in Africa, diseases in which people suffer a high fever and hemorrhages have been reported since 1900—not to mention malaria,* sleeping sickness, yellow fever,* typhus, and the many infectious diseases in Africa that hardly turn a head in the Western world.

On this occasion the world press jumps on the story because westerners are involved—white American women at that. This puts hemorrhagic fever on the map. From now on no one in the wealthy countries can ignore the fact that in Africa diverse forms of this infectious phenomenon exist and that hemorrhagic fever is transmitted from person to person and can kill rapidly. In 1970 it will be shown that the two mission sisters were killed by a new virus that is named Lassa.* This name appears to be a bad idea, because the village will forever bear the stigma attached to a serious disease.

Friday, May 1, 1970
My daughter Ilse is born on Belgium's Labor Day, as Dina, especially, will always recall.

Friday, June 25, 1971
Ilse is over a year old when Dina, with a swollen belly, becomes a doctor in the biochemical sciences. To earn a doctorate, while bringing one child into the world and being pregnant with a second… that demands much mental and physical tenacity. Dina would have preferred to have achieved a doctorate in organic chemistry, but as a woman in 1964 she could not obtain a research assistantship for this. The sophistry behind this was that organic chemistry would be too difficult for a woman, because one must continually lug around heavy flasks of chemicals.

Tuesday, July 6, 1971
The birth of my second daughter, Sonja!

Late 1971
Dina is employed at the National University Center, Antwerp (Rijksuniversitair Centrum Antwerpen, RUCA, later the University of Antwerp), as assistant—in inorganic chemistry—under the supervision of Professor Parmentier.

Six thousand kilometers to the south the "zairising" begins. To wipe out as many traces of colonialism as possible, Joseph Mobutu changes the name Congo to Zaïre and baptizes himself "Mobutu Sese Seko Kuku Ngbendu wa Zabanga," which freely translates to "the almighty warrior, who, thanks to his perseverance and imperturbable will to win, shall achieve conquest after conquest, leaving a trail of fire behind him." The last part of this, especially, is prophetic. The franc is replaced as the monetary unit by the zaïre, the western suit for the *abacost*, and the towns get African names: Leopoldville becomes Kinshasa, Stanleyville becomes Kisangani, and Elizabethville becomes Lubumbashi. Citizens are obliged to change their Catholic names to African ones. In practice many people will go through life with two names, especially in the rural areas. That does not make the identification of people in epidemics easier.

1972

In May a doctor falls ill in Tandala (Zaïre) after he cuts his finger during the autopsy of a youngster. The young man died from a combination of fever and hemorrhage and had been diagnosed with yellow fever. This disease can indeed cause hemorrhagic fever symptoms in some patients. This is still four years before the virus breaks out properly and is named Ebola. Only in February 1977 is the file of the Tandala patient investigated, along with his blood serum.* What happened in 1972, and what occurred elsewhere and earlier—in 1956 in Bili, for example—appear to truly be cases of Ebola before its formal recognition.

In September I am awarded the degree of doctor in biochemical sciences. But first I had to finish my thesis. I originally planned to extract my enzymes from tree bark, but I didn't succeed in making them pure enough to be able to present clear research results. I decided to change my tactic and adapt the subject of my studies to be able to use commercially available enzymes. I was happy to continue making the substrates—the culture media* upon which the enzymes must act. So within a relatively short time I succeeded in mapping the reaction mechanisms of enzymes and finished the thesis that delivered the degree of doctor.

Because Dina works in Antwerp, and the nearby rustic village of Kontich is my birthplace, we decide to buy a small plot there and build a house. At the moment I achieve my doctorate, a permanent position in

the Institute for Hygiene and Epidemiology is offered to me. Simultaneously I receive a proposal to replace Françoise Portaels in the Institute for
Tropical Medicine in Antwerp. She is a doctor in science who is departing
to spend two years in Kinshasa, researching tuberculosis for the WHO.
Antwerp or Brussels? Given two small children and a spouse who works
full time, the decision is rapidly made. I join the ITM,, in "the South,"
and after an investigative discussion I enter the fray as an assistant in the
team of bacteria and virus hunters led by Professor Stefaan Pattyn. For
the biochemist that I am, and who has just a little experience with
bacteria, a whole new world opens.

A SPAGHETTI VIRUS, SOILED SHOES, AND THE BREW OF PÈRE DUBOIS

My inauguration into virology, a blue thermos flask
from the tropics, and the bewilderment of
Flemish mission sisters.

1973–1974

In the microbiology department of ITM, Professor Pattyn assigns me to the study of mycobacteria.* They become my full time business, all the more so because these agents of leprosy and tuberculosis grow very slowly. If I place these bacteria on culture media, I must wait weeks before I can record any research results. Professor Pattyn is also in the thick of it with viruses. After a while I grow bold and ask him if I can apply myself to viruses. Arboviruses, especially, seem a bit more fascinating to me than mycobacteria. Pattyn finds this agreeable. Of course I must get to grips with a whole new field, but that doesn't hold me back.

In the course of 1973 I make another wise decision: I stop smoking.

It's viruses from now on. I breed them in the organs of mice and learn to handle these small animals well. It'll come in handy in days to come. In cooperation with the University of Antwerp, I investigate the effects of drugs on polio, coxsackie, and arboviruses.

During 1974 Peter Piot joins our team, as doctor-epidemiologist. Peter is a couple of years younger than I am, but he will be a brother-in-arms, sounding board, source of inspiration, and a passionate boss.

Tuesday, February 18, 1975

A twenty-year-old Australian tourist dies in a hospital in Johannesburg, South Africa. At the beginning of February the young man was infected in Rhodesia (now Zimbabwe) during an expedition with his girlfriend, during which the couple often slept outdoors. The youngster picked up the Marburg virus (as would later be shown by the CDC in Atlanta). His girlfriend also falls ill, along with the twenty-year-old nurse who cared for the couple. But both young women survive the virus.

August 1975

The German biologist Georges Köhler and the Argentinian biochemist César Milstein succeed in making monoclonal antibodies.* For this they will receive the Nobel Prize for Medicine in 1984. Monoclonal antibodies are an enormous leap forward for the diagnosis of viral diseases.

June 1976

In southern Sudan, not far from the border of Zaïre, a laborer of the Nzara Cotton Manufacturing Factory falls seriously ill. He runs a temperature and suffers severe bleeding: hemorrhagic fever. Nine days later he dies in the overcrowded hospital of Nzara. Meanwhile, another worker at the same cotton factory develops similar symptoms. The man will die eight days later. His wife, who cared for him, suffers the same fate. Less than two weeks later yet another colleague falls ill, one "PG." He is treated for two brief periods in the Nzara Hospital and dies in less than ten days. PG is important for the epidemiological reconstruction of this outbreak. Almost three quarters of the cases in Nzara, and later in two other places, can be traced back to him. According to WHO investigators, his infection was passed on up to six times. The outbreak in southern Sudan is thus severe and will last half a year. In total 284 people are infected and 151 patients die. That yields a case fatality rate (CFR),* or the ratio between the number of cases of a disease and the number of people who die from it, of 53 percent. The cause is a virus that will later be identified as Ebola.

Tuesday, August 17–Sunday, August 22, 1976

In the Equatorial Province of northern Zaïre, a group of people is on an outing. They live in Yambuku, an important village in the Bumba district. Yambuku has a school, a hospital, and a Belgian missionary post that is run by sisters of the Holy Heart of Maria from 's-Gravenwezel and some fathers from Scheut. Among the traveling companions there are the 44-year-old Zairean Mabalo (before Zairisation "Antoine") Lokela. He is the head teacher of Yambuku, next to the border of the Central African Republic. On the way they stop at different villages, among these Abumumbazi, where Mabalo buys antelope meat at the market, to take home. Another Zairean traveler buys fresh monkey. But the company must soon return, because classes begin again in a week in this country, which as a former colony still follows the Belgian school calendar.

Sunday, August 22, 1976

To celebrate the return of her husband, Mbunzu (formerly "Sophie") prepares a banquet with the raw antelope meat that Mabalo has brought. She also cooks the innards. Several family members feast with them. Mbunzu eats the most meat, because she is pregnant. Mbunzu does not eat any of the monkey that another traveler bought.

Thursday, August 26, 1976
It remains to be seen whether Mabalo will be able to stand before a class in a few days, because he doesn't feel well. Most likely it is another attack of his malaria.

Friday, August 27, 1976
Like every other day, hundreds of people come to the hospital at Yambuku, some from far away. From the village of Yaongo, 20 kilometers from Yambuku, arrive Sembo Ndombe and her husband. He requires a hernia operation.

Friday, September 3, 1976
Mabalo reports to the hospital with a temperature and a headache. He is given aspirin and an anti-malarial drug and goes back home. The man from Sembo undergoes his hernia operation.

Saturday, September 4, 1976
Mabalo's symptoms worsen. The teacher now has severe abdominal pain, is nauseous and vomiting. His symptoms are investigated by medical assistant Masangaya Alola, who is also the director of the hospital, where a doctor is only present now and then.

Sunday, September 5, 1976
Mabalo is admitted to the hospital. Nurses do all they can to cure the now deathly ill man. This is clearly not malaria. They try powerful antibiotics, but in vain. Mabalo's vomit turns red and there is also blood in his stools. They are even black from the blood. The condition of the teacher is critical.

Monday, September 6, 1976
Sembo, the wife of the hernia patient, develops the same symptoms as Mabalo. Sembo's husband will later explain that she had had no contact with the teacher or members of his family, but that every day she received an injection at the hospital due to chronic exhaustion.

Wednesday, September 8, 1976
Mabalo loses consciousness. A little later he dies. His body must be made ready for burial. This is traditionally done by women, in this case Mabalo's

mother, mother-in-law, sister-in-law, and some female friends. With their naked hands they remove as much blood and excrement from the body as they can. The ritual lasts 24 hours.

In the hospital at Yambuku there is now a third patient with the same symptoms.

Thursday, September 9, 1976
Sembo dies. The same day two new patients are admitted to the hospital with hemorrhagic fever.

Friday, September 10, 1976
There is once again a new case.

Sunday, September 12, 1976
Yet another new case.

Today is also the hundredth anniversary of the Brussels Geographic Conference that took place on September 12-14, 1876, in the Palace of Laken. King Leopold II had organized the meeting, officially to speak with tropical experts and famous explorers, such as Henry Morton Stanley, about the role of Europe in Africa. But according to many historians, that was only a scientific cloak for Leopold's real ambition: to sate his appetite for a private colony in Africa. All participants in the conference were Europeans and their gathering was the start of the international race to grab African resources. Leopold II recruited Stanley to survey the terrain and prepare the way.

Less than ten years after the meeting, in 1885, private ownership of the Congo Free State by the Belgian monarch was a fact. The terror of Leopold II—with slavery, torture, rape, and mutilation—and the shameless exploitation of ivory and rubber, could begin. It would last until October 18, 1908, the date on which the Congo Free State would be taken over from Leopold II by the Belgian state due to international pressure, becoming the Belgian Congo. In 23 years the population of the Congo Free State had collapsed from 20 million to 8 million, partly through murder but mostly due to sleeping sickness and emigration.

Monday, September 13, 1976
Two new cases of hemorrhagic fever in Yambuku.

Tuesday, September 14, 1976

Four new cases. Not one of the patients admitted after Mabalo were among the company that traveled with Mabalo or ate with him at the feast that Mbunzu prepared. These infections were therefore not due to disease imported from outside the border district. They are all people who have some connection with the hospital or mission, who stay or live there, or come together in the evenings to share a glass. Of the fourteen patients thus far admitted, not one will survive the disease. Among the victims are heavily pregnant women, who enter labor prematurely or miscarry after developing hemorrhagic fever, and one after the other die within the shortest times.

Cases now also crop up in the villages around Yambuku. Most of these patients were in the hospital and returned home.

Until this moment only black people had become sick or died, but now one of the Flemish mission sisters also has a fever: the 42-year-old Sister Beata, born in Mechelen as Jeanne Vertommen. She works in the hospital as a midwife.

Wednesday, September 15–Friday, September 17, 1976

It is now clear: Sister Beata has hemorrhagic fever. On the radio, medical assistant Alola contacts Dr. Mushola Ngoï, medical director of the Bumba district. He makes haste to Yambuku. He is the first doctor there since the outbreak began. By the time he arrives there have been twenty-eight cases, fourteen of which have been fatal. Ten people lie in the hospital with the by-now familiar symptoms. Four patients have fled. Dr Ngoï investigates cases, notes their symptoms, questions family members of the deceased, and describes the hygienic circumstances. He notes all the treatments that the nurses tried in a spirit of despair: fever suppressants, antimalarial drugs, blood clotting drugs, heart stimulants, caffeine, camphor oil... Nothing seems to help.

In the two days that Dr. Ngoï is there, two more patients die. He advises people only to use boiled water, to sterilize medical instruments, and not to bury the deceased next to their houses, as prescribed by tradition. In any case, the doctor is alarmed. He warns his superiors in Kinshasa and writes a detailed and excellent report on his findings—the first of its kind.

Mbunzu, the pregnant wife of Mabalo, is also sick. She develops more or less the same symptoms as her husband, but not the bleeding. Mbunzu ultimately recovers.

Saturday, September 18, 1976
Sister Beata is now so sick that a fellow sister of the nearby mission of Yalosemba comes to offer back-up. Sister Romana, 54-years-old, born in Emblem as Angelina Geerts, is needed because people are still dying in Yambuku. Ever more often the mission sisters contact Kinshasa by radio to report cases of typhus or yellow fever, their suspected diagnoses.

Sunday, September 19, 1976, 5:30 a.m.
Sister Beata dies.

Monday, September 20–Tuesday, September 21, 1976
Dr. Ngoï visits Yambuku again. This time he has brought Dr. Makuta, a colleague from the hospital in the bigger town Bumba, and Dr. Ruranganwa, the company doctor of the Unilever plantations. Dr. Ruranganwa later writes a report that contains the first mention of quarantine regulations that the medical director Dr. Ngoï must enforce at the request of the authorities in Kinshasa. The epidemic will become an economic disaster for Unilever, an important employer in the district. The harvests can no longer be transported and ever more laborers are terrified and flee the region.

Thursday, September 23, 1976
In step with the growing scale of the outbreak, more and more highly positioned functionaries and specialists visit the scene. Dr. Omombo, epidemiologist at the general health directorate in Kinshasa, and Professor Dr. Jean-Jacques Muyembe, microbiologist at the National University, arrive in Yambuku. The intention is that they will spend six days there, but they leave after 24 hours. The scientists find the situation so serious that they want to rush back to Kinshasa to coordinate everything from there. During their short time in the stricken area, Omombo and Muyembe visit a few patients, conduct autopsies, and take blood and tissue samples. The only protective measure that they take is to decontaminate their hands after each activity with iodine and alcohol.

They also have the pluck to evacuate three patients to Kinshasa. The first is Sister Myriam (42, born in Schilde as Louise Ecran), who cared for Sister Beata and who has fallen sick today. The second is Sister Edmonda (56, born in Meerle as Jeanne De Roover), who in turn is nursing Sister Myriam, but isn't yet sick. The third is Father Augustin Claes, who just a month earlier returned from holiday with Mabalo, who became the index case.* It will later become apparent that Father Augustin was never infected.

Saturday, September 25, 1976
After the two scientists and the three evacuees have jolted their way to Bumba in a Land Rover and waited there for 24 hours, they now fly via Kisangani to Kinshasa. At the airport Sisters Myriam and Edmonda and Father Augustin take a taxi to the Ngaliema Clinic, where Sister Myriam is placed in the isolation ward.

Sunday, September 26, 1976
The outbreak looks like it won't stop. Sister Romana, who replaced the sick Sister Beata in the mission hospital of Yambuku, now has the same symptoms. A little later mission pastor Germain Lootens (65, born in Sint-Kruis-Brugge) also falls ill. Father Germain is the hospital chaplain and has already counseled many patients during their battle with death.

Tuesday, September 28, 1976
In the Ngaliema Clinic in Kinshasa, the Belgian doctor Jacques Courteille fills two vials with 5 milliliters of blood drawn from the sick Sister Myriam. He sends the samples in a thermos flask filled with iced water to the ITM in Antwerp.

In the office of the World Health Organization in Brussels a telegram arrives from the medical service of Unilever. It is the first report about what is going on in Zaïre to reach Belgium: "Serious epidemic of an as yet unknown nature prevails in the area of two of our plantations in the Bumba zone, Yambuku Collective. We suspect that possibly a rapidly progressing [form of] typhus or yellow fever is involved."

Wednesday, September 29, 1976
Kinshasa declares Yambuku to be a "danger zone." All traffic to and from the community is forbidden. Forty-three girls are evacuated to Bumba

from the boarding school at Yambuku, just 50 meters from the hospital. The lads had already run away by the middle of September.

In Antwerp, Stefaan Pattyn lets us know that a special package is on its way from Kinshasa for urgent investigation. René Delgadillo, a Bolivian student, and I are curious. Somewhat later a Sabena pilot arrives to deliver the package personally. It is a blue plastic thermos flask of a Chinese make. René and I sit at an ordinary table in the lab, wearing latex hand gloves and lab aprons. Absorbent paper lies on the table, in case we make a mess. To be blunt, these are the protective measures one takes to peel potatoes or clean mussels.

Carefully I screw the lid off the thermos flask. There is iced water inside with a red tint. Beneath the ice blocks there are two glass vials, one of which has burst. There is also a plastic bag in the thermos flask with a folded piece of paper in it. That'll tell us more. I fish the bag out of the water and find a letter from Doctor Courteille. His handwriting says that the vials of blood come from a Sister Myriam who has been struck by an illness that reminds Courteille of yellow fever or typhus. We are intrigued, but it is already late. We carefully place the vials in our freezer. We agree to begin the investigation the next day.

Thursday, September 30, 1976

Sister Myriam dies in Kinshasa after having soldiered through all the appalling symptoms of the disease, and despite treatment with vitamin K1, blood transfusions, and cotrimoxazole,* followed by chloramphenicol,* and penicillin, and even an ultimate attempt with adrenaline and hydrocortisone. Live samples from Sister Myriam are sent to the Mama Yemo Hospital in Kinshasa, whereupon they are immediately sent on to the ITM in Antwerp by order of hospital management.

Meanwhile, we are busy with the blood. Because in Zaïre they suspect a virus to be involved, we start with the procedure to isolate a virus. This goes as follows. First, we grow vero cells* in culture flasks. These are cells that can reproduce endlessly, if you bathe them in a culture medium that contains nutrients. They bind themselves to the wall of the culture flask, and thereby make a thin layer that is just one cell thick. Once the growth phase is over, we introduce quantities of the sample that we wish to investigate—in this case the blood of Sister Myriam—into the cell cultures. The wall of such a culture flask is transparent, so we can observe the cells with a special microscope. The cells are the little factories in which the virus can multiply. Each day more cells are involved, which can deliver millions of viruses after a couple of days. If the cells become sick or die, we call that a cytopathogenic effect.

To be certain that the evildoer is a virus, the experiment is repeated. We harvest the liquid from the cell cultures in which a cytopathogenic effect has been observed, and mix little bits of it with different quantities of growth medium. Thus we make a variety of dilutions. If, for example, you mix one milliliter of cell culture fluid with nine milliliters of growth medium, you have diminished the concentration of the virus ten times. In this manner you make dilutions from 1:10 to 1:1,000,000, or even more. Each dilution is placed in a separate culture flask with new vero cells.

Daily we check each flask under the microscope, to see if there is a cytopathogenic effect. If you see a cytopathogenic effect in the flasks with even the lowest concentration (the highest dilution) after several days, then it is a strong indication that a virus is involved and not some other toxic agent, because the latter would be thinned too much to still make the cells sick or kill them. To check the experiment you also must make flasks of vero cells in the same environment but without the addition of any of the liquid that may contain the virus.

It is a time-consuming and labor-intensive job, and moreover you must work extremely carefully. A prick or cut, a little container that falls over... it happens quickly. Happily I was aided by Peel, a highly trustworthy and careful technician who worked in Kinshasa before Pattyn, in what appear to have been very primitive circumstances. He has given me all sorts of tips. I also pick up a lot from Mrs. Rooyackers, a colleague.

The greatest challenge is to prevent your cell cultures being contaminated by molds or bacteria. If that happens, there is immediately much swearing and we look at each other with an expression of "who did that?" Every contamination means that we must begin again, right from the start. This is work for people with nerves of steel. Moreover, we do everything on an open table and are continually exposed to unsterilized air.

We also inject some of Sister Myriam's blood into the brains of mice, both adults and baby mice. If the animals die, that is also an indication that a disease-causing virus is involved. By means of quick serological tests of Sister Myriam's blood, for which no cell cultures are required, we were quickly able to establish that no yellow fever or Lassa virus was in the samples. The most obvious suppositions were thus ruled out, but more than that we do not know. We must wait for the results of the vero cultures.

Saturday, October 2, 1976
There are two new deadly cases at the mission: at 12:15 Sister Roman dies and at 18:30 Father Germain. In Yandongi and other places in the environs of Yambuku people are dying like flies. Many of them had fled from the Yambuku hospital. Panic in the region reaches its zenith. By telephone and radio Kinshasa gives new instructions to limit the epidemic. No one may enter or leave the Bumba zone. An investigative team will come to check this. Such regulations are not wholly unknown to the local population. Less than twenty years earlier a smallpox epidemic struck the region. Then the people stayed at home, kept their children away from school, cancelled festivities, postponed gatherings, and shut their shops.

In Antwerp, Peter Piot and I go to the laboratory to check the condition of the vero cells and the mice. It is the weekend, but we cannot wait.

Sunday, October 3–Friday, October 8, 1976
The first white doctors arrive in Yambuku: Dr. Jean-François Ruppol, the medical head of the Belgian Cooperative in Zaïre, and his French colleague Gilbert Raffier, chief of the French Medical Mission in Zaïre. Their instruction: to investigate the epidemic on the spot and prevent it spreading to Bumba where two patients with hemorrhagic fever have already been admitted. Ruppol and Raffier stay in Bumba until Wednesday, October 6. They had flown to the area in a C-130 aircraft of President Mobutu and have taken health care equipment and materials from Fométro. Jean-François Ruppol was born in this country and speaks fluent Lingala. That helps.

As soon as they are in Yambuku, Ruppol and Raffier draw blood samples from patients, from people who had contact with patients, and from patients who have recovered. There actually are some survivors. Three. Among them, "Sophie" Mbunzu, the teacher's wife. This is not only fantastic news for Ruppol and Raffier, but perhaps also future patients. A number of the blood samples are sent to the ITM in Antwerp and the Institut Pasteur in Paris.

These are the present conclusions of Ruppol and Raffier:

- The epidemic is much worse than was accepted in Kinshasa following the first reports.
- The focus of the epidemic is the village and mission post of Yambuku, the only hospital within a radius of 50 kilometers.
- The infections are mostly the result of contacts in the hospital and during burials.
- According to the timing with which Sisters Beata, Myriam, and Romana fell ill and died consecutively, the incubation period appears to be eight days, give or take a day, just like the interval between the first symptoms and death.
- Last conclusion: no single treatment helps.

Monday, October 4, 1976

At the ITM everyone is at their post extra early. I throw myself into my work in the laboratory and spot a dead mouse. What about the vero cells? In the microscope I see a cytopathogenic effect in the cells that have been exposed to the blood of Sister Myriam. The control cells have remained normal.

Tuesday, October 5, 1976

Blood samples from Zaïre also arrive at the CDC in Atlanta.

In the ITM a second mouse dies. The supposition that it must indeed be an exceptionally deadly virus grows more probable by the day. It must be said: a virologist is quite excited by this. The kick is not dishonorable and is independent of what happens on the job. It is merely the result of that rare opportunity by which you, as a scientist, can perhaps make a difference in an important aspect of your subject.

But my euphoria is short. The epidemic in Yambuku is spreading at such a scorching pace, the World Health Organization now realizes that the infectious agent must be handled and investigated with the greatest care. The problem is that in Antwerp we do not have a maximum security laboratory (MSL) of biosafety level (BSL)* 4. Two American research institutes have such a lab: the CDC in Atlanta, Georgia, and the military Fort Detrick in Frederick, Maryland. Closer to Antwerp there is one MSL at Porton Down, near Salisbury in England. My boss is not amused when he receives orders from the WHO to send our research material post haste to the nearest BSL-4 lab, Porton Down.

Peter Piot, René Delgadillo, and I must choke back our disappointment while we prepare the infected cells, the brains of the dead mice, and the remains of the blood samples for transport to England. Fortunately Stefaan Pattyn "forgets" to pack a few test tubes with vero cells and a pair of baby mice that are still alive...

Wednesday, October 6, 1976
Some Flemish newspapers report the epidemic. I read the headlines: "Four Flemish missionaries in Zaïre killed by epidemic" and "Secrecy about the Zaïre epidemic—feverish search for unknown virus." And further: "A few days ago the Zairean episcopate was informed by the bishop of Lisala about the fact that an as yet unknown epidemic, caused by an unidentified virus, has killed an untold number of victims, four Belgian missionaries among them. (...) The mission post of Yambuku lies 100 km from Bumba. At first people were of the opinion that typhus was involved, but in spite of the most effective treatments the illness cannot be fought. The victims die just a few days after one has seen the first signs of sickness. More medical teams are being sent to the site with treatment equipment and vaccines. One is groping in the dark as to the precise number of victims, since a great number of inhabitants have left their villages and fled to the forests. One estimates the number at more than a hundred. The mission personnel in the mission post of Yambuku, which was founded in 1936, comprised three Scheutists, six sisters of the Holy Heart of Maria, 's Gravenwezel, and a Zairean brother."

In the lab I fix a number of the infected vero cells that we still have with formalin. If there was still a living virus, now it is certainly dead. These are cells that I wish to investigate with the electron microscope. We don't have one ourselves, but I can entrust the sterilized preparation to Wim Jacobs of the University of Antwerp and on October 11 we will have the results.

While we are still busy in the lab with the other vero cells, Stefaan Pattyn enters. It isn't often these days that he wishes to crouch behind the microscope himself—so we see just how curious he is. Pattyn walks to the rack where we keep test tubes and grasps one, but it slips out of his hand! René and I see the tube shatter into glass shards on the floor. Part of the contents lands on our shoes. After swearing vehemently, we clean the floor and our shoes as rapidly as possible with paper and hypochlorite disinfectant. Not much could have happened, but we are not altogether sure if a virus is involved, and if so, how that virus spreads itself. Perhaps a droplet did come down on one of our heads and landed in the area of a tiny wound. In any case, Pattyn seeks us out every day to take our temperatures and orders us to report the slightest fever.

In the lab a few more baby mice die. We have also received the liver sample of Sister Myriam. This specimen, which is fixed in formalin, is immediately investigated by our colleague Paul Gigase, senior lecturer and an experienced pathologist and anatomist. He determines that a type of tissue damage is present that is compatible with infection by a hemorrhagic fever virus.

Thursday, October 7, 1976
At ITM Stefaan Pattyn prepares a report for Fométro in Kinshasa, in which he states that he received blood samples on the evening of September 29, that the laboratory staff began to work with the specimens on September 30, that a virus is probably involved, but not Lassa or yellow fever, and that the samples will be investigated on October 11 with the power of the University of Antwerp's electron microscope.

In the lab we find one dead, one paralyzed, and one sick baby mouse. We remove the little brains and send them to Porton Down in England.

Today the newspapers write about the epidemic again: "Unknown disease in Zaïre not yet contained—President Mobutu becomes personally involved." I read: "On Wednesday the sisters of the Holy Heart of Maria of 's Gravenwezel learned from a report of the Scheutists that the fathers and sisters were evacuated by helicopters from the center of the stricken region. Also at the Ministry of Foreign Affairs people are now more involved. So we learn that Dr. Ruppol, head of medical cooperation between Belgium and Zaïre, on the personal request of President Mobutu, left on a military flight from Kinshasa to Yambuku to lead the fight against the disease there. (...) It is certainly not an exaggeration to speak of a disaster. The World Health Organization is also now involved. They warn of the highest degree of infectivity and of the exceptional danger of an unknown hemorrhagic fever, caused by a virus. Therefore it is likely that extensive samples of infected blood were sent to the Institute for Tropical Medicine in Antwerp, to the Institut Pasteur in Paris, to the university research centers in Great Britain and the United States. (...) In the various laboratories staff had by Wednesday evening not yet made progress in the task of identifying the unknown virus, about which some people have perhaps been all too premature to discuss opinions that it is Lassa fever. All experts and their colleagues are working ceaselessly to ring-fence the catastrophe but first they must discover the cause of the disease." Hey, that's motivation for you!

Friday, October 8, 1976
In the Ngaliema hospital in Kinshasa, Sister Edmondo falls ill.

Monday, October 11, 1976
In Antwerp, Wim Jacobs of the Univerity of Antwerp has examined our preparations under an electron microscope. The photographs of the virus magnified ten to twelve thousand-fold are shocking. What a giant!—a threadlike giant. It resembles a spaghetti noodle. Or a large worm, thinks Peter Piot. Some of the entangled viruses are a good 14,000 nm long. That is only 14 millionths of a meter, but is nonetheless enormous in comparison with the capsule of the polio virus, which is 140 times smaller. But is it also new? As Pattyn studies the photos, he notes that our leviathan significantly resembles the virus that was isolated in Marburg in 1967. Regardless, we now know for the first time that the Yambuku epidemic is caused by a virus, possibly a new species, and that the agent of the disease is morphologically related to the Marburg virus.

In Paris, at Roissy Airport, Pierre Sureau, a respected expert and researcher at the Institut Pasteur, has come to collect samples that were sent that day for him by his compatriot Gilbert Raffier, from Kinshasa. But the samples must be found. Sureau understands that the specimens were given to a pilot in Kinshasa. He took them with him to France. Pierre tracks down the number of the flight commander and phones him. The man says that the army medical service immediately took charge of the package upon its arrival at Roissy and that it is now in the military hospital of Percy in Clamart, to the southwest of Paris.

After many detours Pierre can finally open the mysterious package in the late afternoon. Just as I had two weeks earlier, he finds a thermos flask with two blood samples (these from Sister Edmonda) and a short accompanying letter. And just as in Antwerp, there is an incident. The two test tubes are cooled with dry ice, but now that they are suddenly exposed to room temperature, one of them breaks.

A little later Pierre receives a telephone call from his friend Dr. Paul Brès, serving head of viral diseases at the WHO in Geneva. Brès announces that Pierre will receive a little package from Zaïre in which there are possibly highly infectious samples that may only be handled in a maximum security laboratory. He requests that Pierre, above all else, does not open the package and sends it at once to the CDC in Atlanta. When Pierre says that it is already too late, Dr. Brès switches immediately to his second request. He explains that the WHO has just been informed of an outbreak of a highly infectious hemorrhagic fever in Zaïre, of unknown cause, and would value it greatly if Pierre would be prepared to travel directly to Zaïre, as the World Health Organization's consultant. Pierre already has experience: in 1972 and 1974 he conducted epidemiological investigations in the Bangui division of the Pasteur Institut, Central African Republic. Amongst others, he worked there with the notorious Crimean-Congo Hemorrhagic Fever Virus.

Pierre doesn't need to ponder long over the offer. Here in Paris he is becoming a little bored after his taste of fieldwork and adventure in Africa. Paul Brès phones Pierre Sureau's director to rapidly arrange this and that. Pierre drops everything and spends the rest of the day preparing for his expedition.

Tuesday, October 12, 1976
Pierre Sureau finishes preparing the dangerous packet from Zaïre—extra secure this time—for dispatch to Dr. Karl Johnson, head of the Special Pathogens Branch (the division of special disease-causing agents) in the CDC and discoverer of the Machupo virus. His flight to Kinshasa leaves at 21:40. When he reports to the Swissair check-in with his baggage and medical equipment, the check-in clerk warns him that it's dangerous to travel to Zaïre because an epidemic has broken out there.

Wednesday, October 13, 1976
At 07:00 Pierre is collected at Kinshasa by Dr. Gilbert Raffier, Dr. Collas of the WHO, and Dr. Jean-François Ruppol. Pierre had no time to arrange a visa but Jean-François Ruppol knows who has the upper hand here and pilots the Frenchman through the police and customs. At last he is again in Africa.

Pierre Sureau is the kind of epidemiologist who prefers to take action immediately. This is also essential, because in the WHO office a newly arrived telex is waiting for him with the request to send new blood samples as quickly as possible to the CDC in America. Pierre wants to go the hospital right away, but he first must formally announce his presence to the state secretary for health, Dr. Ngwete Kikhela, who proposes that he travel as fast as possible to Yambuku, although that will only be possible on Saturday due to organizational problems.

When Pierre eventually reaches Ngaliema Hospital at 17:00, after all sorts of other duties, he at once visits Sister Edmonda. The 56-year-old sister accompanied Dr. Muyembe from Yambuku three weeks earlier and at the moment she is the only patient in the Zairean capital, as far as is known. Pierre dons the protective clothing that has only been available in the hospital for a few days: a paper apron, a hair net, gloves, glasses, and a mouth mask. He draws blood from the sister and tries to encourage her. But Edmonda, who has lost two of her fellow sisters and numerous Zairean friends, has no illusions. She has lost a lot of weight and knows what awaits her. Pierre is impressed by the strength that she radiates.

It is already pitch dark when Pierre walks to the garden of the hospital and there sets his protective garb aflame. He doesn't yet know that at that instant in the very same hospital Mayinga N'Seka, a Zairean nurse, has developed the notorious symptoms. From September 27–30 Mayinga cared for Sister Myriam who lay here in isolation. She did this task together with Sister Edmonda. Protective clothing wasn't available then.

In Yambuku the nuns leave the big house where they live because their fellow sisters Beata and Romana died there. They lock themselves into the mission's guesthouse and rope off the area around their new ward with long, white bandages—a literal *cordon sanitaire*—which they fasten to the shrubs and some poles. They hang a sign up with the text: "No entry. To call the sisters, ring the bell."

Thursday, October 14, 1976

Just as the morning of what will be a very busy day begins—in Kinshasa, Yambuku, Johannesburg, Atlanta, Antwerp, and Geneva—Sister Edmonda passes away. Because there were strong suspicions that the cause of her disease was viral, she had received the antiviral drug Virustat but didn't benefit from it. Pierre is knocked off his feet when he learns that Sister Edmonda has died. The mission sister leaves a prayer as her deathbed gift: "My good Zairean friends of Yambuku, a treacherous epidemic weighs heavily upon you with its deadly effects. May the Lord accept the offer of my life, for the well-being of you all."

In the small lab of Fométro, Pierre Sureau finishes packing the blood sample that he drew the previous evening from Sister Edmonda. He sends the specimen via New York to Karl Johnson in Atlanta. Meanwhile, the report written by Professor Pattyn on October 7 arrives at Fométro. Pierre has almost finished reading it when a telex chatters in, also from Stefan Pattyn, and he finds this report more interesting. Pierre reads that after studies with an electron microscope a Marburg-like virus has been isolated in Antwerp. For Pierre, it is big news. He recalls the cases of Marburg in South Africa a year earlier—the student couple. Then, one patient died but two recovered. With the serum of the survivors he might be able to save Sister Mayinga. Because Zaïre has no diplomatic relations with the apartheid regime of South Africa, Pierre ropes in Dr. Bill Close. As the personal physician of President Mobutu, Bill can pull on a couple of strings. His trouble yields results, because a few telephone calls later Dr. Margaretha Isaacson, head of epidemiology at the South African Institute for Medical Research in Johannesburg, is readying herself to depart for Kinshasa with the serum.

In Zaïre time is beating the drum, because Father Augustin, back in Yambuku after his stay in Kinshasa, lets Fométro know via radio that in the meantime there have been 253 cases within a radius of 20 kilometers around the mission post: 99 men, 97 women, 24 children older than 5 years, and 33 children younger than 5 years. Today the medical assistant, "Sophie" Mbunzu's mother, and two other people have fallen sick. Yambuku hospital is closed.

On the same day Karl Johnson and two other CDC virologists, Patricia Webb and Fred Murphy, show that the virus we have seen in Antwerp is of a different type to Marburg. To be specific, the samples that the CDC has received from Zaïre do not react with antibodies against the Marburg virus. This is not any form of Marburg. Scientists at Porton Down in Britain come to the same conclusion. The scientists in America and England immediately dispatch their results to the WHO in Geneva. From there Paul Brès sends the following telex to the world: "On 14 October

46

1976 the ITM (Antwerp), Belgium, Porton Down (London) [sic], Great Britain, and the CDC (Atlanta), US, simultaneously isolated a virus that is morphologically related to the Marburg virus but immunologically different."

We have discovered a new virus!

Friday, October 15, 1976

The WHO lets its African affiliate in Brazzaville know that the CDC wants to send a team to Zaïre, in view of the seriousness of the epidemic. This is sensitive, because although Mobutu has an American personal doctor, the Zairean public health services are not keen on Western aid, and especially not if it comes from the United States.

Also in Belgium the administration quietly becomes convinced of the gravity of the situation. There is mild panic, so they hear, among the Belgian workers in Kinshasa. The government is under pressure to do something. The first phone calls that Professor Pattyn receives this morning come from the Department of Development Aid and the Ministry of Foreign Affairs. There must be rapid intervention. The administration requests Pattyn to put a team together, that can leave for Kinshasa in two days, to join an international medical commission there. As an epidemiologist, Peter Piot must go, but right now he is attending a conference in Paris. Pattyn rings Peter and asks him to return without the slightest delay. I am also ordered to travel to Zaïre by Pattyn, but only in a week's time. Due to the incident with the test tube and the shoes it is not yet clear that I am healthy. If I remain fever-free, I can leave on October 22 at the earliest. For both Peter and me it will be our first trip to Sub-Saharan Africa. Pattyn expects that we will stay there for roughly fourteen days.

In the epidemiological journal of the WHO, two pages appear concerning the current epidemic and earlier outbreaks of hemorrhagic fever. This is not only the first official media report about the Zairean epidemic, but also about the outbreak in the Sudan, about which only a little has been known until now.

In Flanders, the *Gazet van Antwerpen* headlines read: "African virus identified—Another Belgian mission sister dies." I read: "The cause of the mysterious deadly epidemic in Zaïre and Sudan could be determined after intensive research. It is a variant of the so-called Marburg virus, feared in Africa as 'green monkey disease.' (…) There is obviously great rejoicing at the discovery of the virus, in particular at the Tropical Institute in Antwerp where staff have nerve-wracking weeks behind their backs. The virus was sought not only in Antwerp but also in the United States and England. The virus could be identified in the three locations thanks to an electron microscope. But the rejoicing is still overshadowed by the fact that as yet it has been impossible to find antibodies to fight the disease. One hopes to be able to do this quickly as was the case for the Marburg virus. (…) Prof. Pattyn says (…) that a principal line of attack is to find out where the so-called 'virus reservoir' is located and how it is transmitted to humans."

In the evening, Pierre Sureau learns that Karl Johnson and Joel Breman of the CDC are ready to leave for Geneva. There they will await the green light for Kinshasa. Although we have found a new virus, the word "Marburg" has dropped and the horror that this word evokes washes everything into the rapids.

Saturday, October 16, 1976

Dr. Margaretha Isaacson arrives in Kinshasa at 01:00 in the morning from South Africa, armed with the Marburg serum.

At the ITM we have come together on this Saturday morning to make the last preparations before the departure of Stefaan and Peter. We assemble work and protective equipment. Peter receives a cram-course in maximum security and laboratory procedures. I give him motorbike goggles, which protect both his eyes and nose well.

In Kinshasa, Pierre hears that his departure for Yambuku—planned for today—is again postponed. He profits from his unexpected time in the capital by including extra protective equipment: one hundred refuse bags of strong plastic that can double as disposable boots, a dozen blue overalls, cheap plastic motorbike goggles, and fifty flasks of bleach.

For the first time the Zairean press bureau distributes a report about the epidemic. It states that the Marburg virus has broken out within a radius of 20 kilometers of Yambuku, but that the epidemic is now subsiding.

In the Ngaliema Hospital, all health workers of ward 5 are quarantined after the admission of nurse Mayinga. Pierre takes 22 blood samples from them and also from the members of the international medical commission. After the weekend these samples will be sent to Atlanta.

Pierre spends the rest of the day with Margaretha Isaacson, next to the bed of Mayinga. At 22:00 Mayinga is given 100 milliliters of Marburg serum. Pierre and Margaretha know that the Yambuku virus is no Marburg, but it resembles it and they have nothing better. The serum is Mayinga's only hope. Pierre stays at her bedside through the greater part of the night. Every two hours he draws a little blood.

In Yambuku it appears that the epidemic reached its peak a couple of days ago, but they are certainly not yet convinced. The housekeeper of the school principal dies. A young pupil of the mission school also dies.

Sunday, October 17, 1976

Mayinga now shows the typical rash on the trunk that many patients in Yambuku also had. Due to severe throat pain she can no longer eat or drink by mouth and must do this through a drip. Jean-François Ruppol goes to the center of Kinshasa to find popsicles to alleviate her pain but finds that not a single shop is open. The doctors therefore let Mayinga suck ice cubes and give her a second dose of the Marburg serum. In the evening Stefaan Pattyn and Peter Piot catch the flight to Kinshasa.

Monday, October 18, 1976

René Delgadillo and I turn to the cultures that still stand in the lab and finish the harvest: we divide the cellular material into small tubes that we freeze to create a small store of the virus. I do the same with the tissue of the dead mice. After this we clean everything thoroughly.

Karl Johnson and Joel Breman of the CDC arrive in Kinshasa. Bill Close went to fetch his compatriots in Geneva and during the flight to Zaïre gave them an expansive account of the situation in Kinshasa and Yambuku. He is able to push the Americans quickly through. Karl and Joel have brought a fluorescence microscope, an important item, and more: a centrifuge, a portable isolator in flexible plastic for the manipulation of dangerous samples, containers with liquid nitrogen, and many types of smaller lab equipment. Karl, who is *au fait* with the highest level of practice in virology, immediately suggests that he leads the international commission, and this isn't contested. The commission will comprise two groups, one in Kinshasa and one in Yambuku. In each group people are allocated to patient treatment, laboratory activities, and epidemiology.

At 17:00 the first meeting of the international commission takes place in the meeting hall of Fométro. The 22 present are, grouped by nationality:

- Dr. Ngwete Kikhela (State Secretary for Health, Zaïre)
- Drs. M. Koth and S. Matunda-Nzita (City Hygiene, Kinshasa)
- Dr. Tshimbamba (City Inspectorate, Kinshasa)
- Dr. Madiangu (Fonds National Médico-Sanitaire, Kinshasa)
- Dr. Omombo
- Dr. Lekie
- Drs. René Collas, L. Djordevic, and S. Adrien (WHO, Kinshasa)
- Dr. Pierre Sureau (Institut Pasteur, Paris, France, and WHO consultant)
- Dr. Gilbert Raffier (head of the French medical mission in Kinshasa)
- Dr. Margaretha Isaacson (South African Institute for Medical Research)
- Dr. William Close (medical service of the president, Zaïre)
- Dr. Karl M. Johnson (head of special pathogens, CDC, Atlanta, US)
- Dr. Joel Breman (medical epidemiologist, CDC, Atlanta, US)
- Dr. John Kennedy (head of public health, US Aid, Kinshasa)
- Prof. Dr. Stefaan Pattyn (ITM, Antwerp, Belgium)
- Dr. Peter Piot (ITM, Antwerp, Belgium)
- Dr. J.F. Ruppol (head of the Belgian medical mission in Zaïre)
- Dr. Jean Burke (head of Fométro and medical adjunct of ABOS, Brussels, Belgium)
- Dr. D. Thonon (Belgian medical cooperation, Kinshasa)

In the Kinshasa group the Danish-Dutch–South African Margaretha Isaacson will focus on therapeutic aspects, the American Joel Breman on epidemiology, the Frenchman Pierre Sureau on virology, and the Belgian Stefaan Pattyn on laboratory work. It could not be more international. Dr. Ruppol will manage the administrative and organizational aspects. In conclusion, a team is put together to travel to Bumba and Yambuku the next day to assess the situation, to check if and how a treatment center and lab can be organized, and to draw blood samples from recovered patients. The team consists of five doctors: Koth, Breman, Piot, Ruppol, and Sureau. A last instruction for the gathering: Karl Johnson requests all members to take their temperature twice per day. Later it will be seen that Pierre Sureau, Peter Piot, and I fall ill independently, and not one of us obeys.

In the early evening it seems that the fever of nurse Mayinga has fallen to 37.5°C (99.5°F) following two doses of serum containing antibodies against Marburg. Margaretha Isaacson draws up a list of recommendations to prevent the further spread of the virus in the Ngaliema Hospital.

Tuesday, October 19, 1976

Pierre Sureau is glad that he can finally travel to the epidemic's hearth, 1,200 kilometers from Kinshasa. After breakfast he gets a few more pointers from Karl Johnson and checks if the containers of dry ice and the protective and lab equipment have been loaded on the four-by-four. At about 11:00 the C-130 takes off with the five doctors of the Yambuku team on board. Underway and sitting on uncomfortable little seats, they read a book about the first Marburg outbreak in 1967.

In Kinshasa, nurse Mayinga is again sicker. She vomits blood and is seriously weak.

After a short stop in Mbandaka, the C-130 lands at about 14:00 in Bumba. The Belgian Scheutist Father Carlos welcomes the company. They obtain an oversight of the course of the epidemic. The doctors spend the night in the mission post, after a tasty meal and many Primus beers on the mission veranda.

Wednesday, October 20, 1976

After breakfast Sureau departs for the isolation camp in Bumba. There he draws blood samples from some people who have been in contact with two patients who died from hemorrhagic fever in Bumba. Dr. Massamba Matondo, the doctor of the Lisala subdistrict who has visited 44 villages in the area, tells Pierre that all infected people were in Yambuku, or had contact with people who had fled from Yambuku.

Pierre and the others ready the documents they will need to enter the closed zone. The group of five, supplemented by Dr. Matondo, leaves for Yambuku in two Land Rovers—one they brought with them from Kinshasa and one they borrow from Father Carlos. During the trip of 100 kilometers over abominable roads, they check the roadblocks while probing for possible patients and gauging the movements of people.

After five hours driving the team arrives in Yambuku. They are greeted from a distance by a father, three sisters, and two Zairean nurses.

They look happy to see the team arrive but do not dare let the commission's members come close. Peter stoops under the bandages of the *cordon sanitaire* that the sisters have set up. The others follow and the team makes its acquaintance with Father Leo Claes (also known as Père Dubois), Head Sister Marcella (Julienne Ronsmans), Sister Genoveva (Annie Ghysebrechts), Sister Mariette (Maria Joanna Witvrouwen), and two young nurses from Lisala who have been here for about ten days, blocked by the quarantine.

Pierre is dog-tired, but that same evening he has Sister Marcella give him the full picture of the situation. She gives him a carbon copy of her typewritten notes. From her scrupulous report, it appears that within one month 38 of the 300 people at the Catholic Mission of Yambuku had died, among which one of the three Belgian missionaries, three of the six Belgian sisters, the Zairean medical assistant, four of the six Zairean nurses, and one of the two cleaners.

In the months before the outbreak no medical consultations, fever cases, surgical interventions, miscarriages, or deaths out of the normal were noted, although in August there were three cases of fatal bleeding following a birth.

Sister Marcella also names the nine villages where the latest news has it that there are still sick people. Since the quarantine, the sisters no longer make the rounds of the villages. So that's why there are no more sick people in Yambuku, Pierre thinks; it's because they have fled to other villages where perhaps they are spreading the disease farther. Despite his interest in learning more from Sister Marcella's detailed document, Pierre goes to sleep with his colleagues in the abandoned girls' dormitory of the boarding school, with more questions than answers.

In Kinshasa, nurse Mayinga passes away after a painful struggle against death.

Thursday, October 21, 1976
The first epidemiological investigation in the region around Yambuku takes place. Transport is crucial. Four all-terrain vehicles are available: two from the mission itself, and the two Land Rovers the team brought from Kinshasa and Bumba. Four pairs divide the area along as many axes, and thus they will interview more than ten thousand people and track

down sick and recovered patients. To limit the risk of infection, only the fourth pair—Peter Piot and Pierre Sureau—take blood samples from sick patients that they or the others have found, and possibly also from recovered people or people who were not sick but who have had close contact with patients.

Sister Marcella knows the region like the back of her hand and will drive Peter and Pierre around in the Land Rover that she normally uses to tour the villages. She is very happy that after one week her isolation is over.

Early in the morning in Yalikonde, Peter and Pierre meet their first sick person: Lisangi Mobago, a man of 25 who comes from Yaombo, where on October 8 he attended the burial of someone who had succumbed to the "Yambuku disease." Six days later Lisangi became sick and now he has arrived in Yalikonde to consult the local healer.

Later, at 09:00, the two researchers are led to a hut in Yamolembia where they encounter a deathly ill young man of 22. He dies under their eyes, even before Pierre could draw blood. Pierre makes no further attempt, out of fear that his neighbors will connect the sampling of blood with the death.

Peter, Pierre, and Marcella travel the whole day from village to village and patient to patient, some being much sicker than the others. Now and then they can take blood samples from people who possibly had the disease. Maybe they will deliver serum with which acute patients can be treated.

In the late afternoon a thunderstorm floods the sand roads, exacerbating the return journey to the mission. Due to the storm, the helicopter that the team was relying on to visit a number of villages that cannot be visited by road, doesn't reach Yambuku.

In the evening the report of Mayinga's death arrives by field radio. Sister Marcella and Pierre cannot hold their tears back. Mayinga was a beautiful young woman with her whole life ahead of her, who planned to study abroad. When Pierre left on Wednesday, she looked as if she was gaining the upperhand, but the serum that Margaretha Isaacson brought from Johannesburg, the best care, the nights that Margaretha and Pierre spent at her bedside... it had all been in vain. To combat the somber thoughts, the sisters roast a sheep. Father Léo "Dubois" lets the company make acquaintance with his homemade cocktail of vermouth, banana alcohol, and orange juice. Pierre realizes the origin of the nickname "Père Dubois"—just like two peas in a pod, the father resembles the face that appears on the label of a bottle of Elixir du Père Dubois. The valiant father is proud that he hasn't gone down with hemorrhagic fever and ascribes this to the fact that every day he has drunk a shot of his own concoction.

VOICES IN THE GARDEN, A HEADLESS SNAKE, AND THE SPECTRE OF BIOLOGICAL WARFARE

My first activities in Zaïre, the United States, the Soviet Union, and Japan. And how I build my own maximum security lab and go in search of Ebola's elusive host.

Friday, October 22, 1976

My mother turns 62, but I can't be at the party. After the incident with the test tube in the ITM lab I measured my temperature regularly and did not develop a fever. Thus I was not infected, and with my mind at rest I can depart for Zaïre, for the first time in my life. I do not need to spend much time preparing. The research equipment is already in Africa, my passport is in order, and I don't need much luggage for two weeks.

In Yambuku, the international team discusses the activities of the previous day. Peter Piot, Pierre Sureau, and Sister Marcella have seen many acute patients. They have noted that not all these patients are wholly isolated—some maintain direct contact with their surroundings. In other places, sick people must indeed stay indoors—their food is set down by the hut entrance and they are otherwise left alone, to the bitter end. After this the hut is burned.

In the western zone, Joel Breman and Father Léo (Père Dubois) have visited seventeen villages. The epidemic has struck in sixteen of these.

The team that has visited the south and east found no acute cases. But in Yamakondo they learned that of 11 people who had been in the hospital at Yambuku, 10 have died of hemorrhagic fever. Also in the other villages it appears that virtually all patients had been in Yambuku.

To conclude, Jean-François Ruppol in the northern sector announces that the quarantine regulations have worked well for his area.

At 08:00 the commission members visit the hospital, which has now been closed for a whole week. A nasty atmosphere hangs over the abandoned dispensary. In the maternity unit the doctors find flagons of antibiotics and other drugs, the rubber stoppers of which have been perforated or replaced by plasters. There are also plastic syringes scattered around. With these, tens and sometimes hundreds of injections were given daily—not only to the patients admitted, but also to the long rows of sick patients who came here daily for treatment as outpatients. Without an injection you hadn't received treatment—well, not a proper treatment according to the mentality of the people here at one time. Once each morning the needles were sterilized, and never again the rest of that day. Later some of the sisters will deny this and claim that the needles were sterilized more than twice per day. Whatever the truth of this, the epidemiological data show that the hospital was the most important source of infection, even for people who just passed through and never had contact with hemorrhagic fever patients.

In the afternoon, Pierre receives a radio report from Karl Johnson in the capital. Can Pierre bring Sophie (Mbunzu) to Kinshasa for plasmapheresis?* Mbunzu, the wife of the deceased teacher, is one of the few patients thus far who is known to have recovered. Her blood possibly contains antibodies that can save others.

After midday, Pierre draws blood samples from Dr. Matondo, the doctor of the Lisala subdistrict, and from nurse Mbongo Ebanga, Père Dubois, and the sisters Marcella, Genoveva, and Mariette. A week later in Kinshasa I will find antibodies in these samples.

Pierre and Sister Genoveva convince Sophie to accompany them to Kinshasa, together with her baby. They also recruit Sukato Mandzomba, a 24-year-old male nurse who became sick on September 21 but who has recovered, just like Sophie. The same day, Pierre, Peter, Joel, Sukato, Sophie, and her baby drive to Bumba, to fly as soon as possible from there to Kinshasa.

I arrive in Kinshasa in the late afternoon after a smooth flight. The animation of the city is impressive. Nothing here is gray or drab, everything jars exuberantly. The streets are a gaudy, frivolous ants' nest of wriggling traffic. Everyone drives how and where he wishes. A policeman stands alongside and watches it all.

I find shelter in a Fométro building. It is almost time to unpack my things, because I am expected at a briefing. There I become acquainted with the captivating team leader, Karl Johnson. After a while I will frequently call him "Papa Machupo," after the virus that he discovered in Bolivia. The information that Karl gives me about our virus is highly alarming. The virus is new, deadly—albeit not 100%—and we know little about how it is transmitted. There is tension among the team members who have stayed in Kinshasa. After the death of Sister Mayinga they are terrified that there will be other cases in this steaming metropolis of two million residents. The nurse had contact with a considerable number of people. For example, she went to various ministerial departments because she wished to travel to the United States for supplementary training. Happily, Mayinga was extremely professional. As soon as the first symptoms appeared, she made a list of the places where she had been and the people she had met. Thus the "at risk contacts" had been admitted to the quarantine block of the Ngaliema Hospital.

I learn that I am to establish a small, well-secured laboratory inside the hospital. There, with the help of indirect immunofluorescence, I can detect antibodies against the new virus in the serum of infected patients. For such a test you need a UV microscope and glass slides upon which there are fixed vero cells with the killed virus. Karl Johnson and Joel Breman have brought the essentials with them from the Center for Disease Control in Atlanta. Together with Patricia Webb they hurriedly made glass slides with the dead virus just before they left. The virus had only become available in their lab barely days before. There was not enough time to check that the fixed virus really was dead, So Karl advises me to be extra careful.

Saturday, October 23, 1976
Today and for the whole of the next week I am in the trenches, constructing and organizing a laboratory where I can work as safely as possible with a lethal virus.

Sunday, October 24, 1976
In Bumba it is a day of celebration. After a month of isolation the quarantine imposed on the forty schoolgirls who were evacuated from Yambuku has expired.

Peter and Pierre hear from Kinshasa that two Cessna light aircraft have been arranged to fly them with the recovered patients Mbunzu and Sukato to Kinshasa tomorrow. But the Cessnas might still become a C-130 Hercules or arrive a little later in the week, if no fuel can be had or the weather is too bad.

Monday, October 25, 1976
In Bumba, Pierre and Peter's little group constantly awaits an aircraft. However, they learn there is a helicopter available—and not just any old helicopter, but the private Puma of Mobutu! At that moment the president is staying in Switzerland and doesn't need it. The Puma carries them back to Yambuku, where they can do a little more work while waiting for their flight to Kinshasa.

When they land at the mission, Jean-François Ruppol has just finished with a difficult birth. But the children of the village only have eyes for the Puma. In the wink of an eye a pair of them have fashioned a toy version of the alien metal bird from a few scraps of wood and some nails.

Nothing more was heard that day about the arrival of an aircraft in Bumba. It is especially worrying that the blood samples Pierre collected in Yambuku and the surrounding areas threaten to dry out as they wait. Once they are back in Bumba, Pierre drives with Peter to the building of Unilever in Ebonda, 15 kilometers farther on. There they have a large deep freezer that always works. The director grumbles when he hears that he must keep a virological bomb cold, but he eventually consents.

At the office of the United Nations in Kinshasa, a telex arrives with news about the other epidemic in southern Sudan, with the report that three laboratories in Antwerp, Atlanta, and Porton Down have discovered a new virus that resembles Marburg, but which has different antigens.*

It rapidly becomes apparent that I cannot count on the help of the Zaireans to set up my lab. Since two Belgian mission sisters and a Zairean nurse passed away in the Ngaliema Hospital, panic has reigned. Also outside the hospital people have heard about the deadly disease. It was in the newspaper. Residents receive the advice to avoid Ngaliema and especially that part where patients are held in quarantine. Everyone I ask to lend a hand with my laboratory gives me a friendly but firm refusal.

Tuesday, October 26, 1976
In the morning the fathers in Bumba once again learn via the radio that there is a C-130 in the pipeline to carry Peter, Pierre, and the others to Kinshasa. When is not known, therefore they wait. Everyone tries to keep himself busy. Reports are written, siestas slept, discussions held with the fathers. In Bumba it rains the entire morning. There is no sign of an aircraft.

In Kinshasa, I find a few members of the international commission ready to help haul a few concrete blocks and items of furniture to my lab in ward 5.

Wednesday, October 27, 1976
A radio report from Kinshasa: the C-130 is now truly on its way. Pierre Sureau hurries to Ebonda to fetch his blood samples from the deep freezer.

At 10:30 the four-engined colossus rumbles down onto the landing strip in a cloud of red dust. As the little group of passengers stand ready, it seems Joel Breman is nowhere to be found. Father Carlos and Pierre tear back to the mission with the Land Rover. No Joel. His belongings lie spread over his bed. They pack and grab his luggage and hurry back to the little airfield. There they find Joel coolly waiting for them. He had gone to take photos.

Upon boarding, things threaten to go awry once again. The flight commander doesn't want to allow the two "patients" and the blood samples on board, notwithstanding that they are—note well—the very reason he had to fly there with his C-130. After some difficult bargaining and some smart *matabiche** everyone may board. At 11:15 the uncomfortable cargo plane takes off. This is Sophie and Sukato's baptismal flight. They will endure it without a squeak. At 13:15 the aircraft lands in the capital and they are at once taken to the Ngaliema Hospital.

This evening the international team is united and at full strength for the first time. In Fométro's communal hall it is pleasantly crowded. Fortunately, after the death of Mayinga eight days ago, no new cases of hemorrhagic fever have appeared, and also none elsewhere in Kinshasa. Tales are swapped. Everyone is thinking of Yambuku and wants to go there as soon as possible.

Thursday, October 28, 1976

Together with Karl Johnson, I investigate the specimens that Peter and Pierre brought with them from Bumba, and of course also the sera of Sukato Mandzomba and Sophie Mbunzu. We test for both Lassa and the new hemorrhagic fever virus, which still doesn't have a name. Karl teaches me to work in an isolator. In America he manages a maximum security lab and I make full use of his experience. Only at 01:00 have we finished and we know three important things for sure. One: no one has contracted Lassa; two: Sukato and Sophie do indeed have antibodies against the new virus; and—a sigh of relief—three: the sisters, Père Dubois, and Dr. Matondo are not infected, which proves that you can mix very closely with lethally infected people without being stricken yourself.

It is a historic day. For the first time a diagnostic test has been conducted that indicates if someone is actually not infected with the new virus. It is also the first serological proof of the clinical phenomena that have now been recorded in the field for more than one and a half months.

Friday, October 29, 1976
After having visited Sophie and Sukato, Pierre Sureau comes to see what I have cooked up in my "virological laboratory." Appreciatively, he inspects the isolator of flexible plastic in which I handle my research material. The construction is built up on a table that I raised with concrete blocks. I work as carefully as possible and wear the protective clothing at my disposal. The fluorescence microscope stands in the bathroom, because it is the only space I can darken enough to see my "little stars"—this is what infected cells look like, under ultraviolet light. That I could improvise a safe work environment with simple means is praised by Pierre—if not the first, the best, by far. His words do me much good.

But there is also bad news: there is a rumor that hemorrhagic fever has broken out in a prison in Kikwit, a town 350 kilometers south of Kinshasa.

Saturday, October 30, 1976
The epidemiologists Joe McCormick of the CDC and the Belgian Simon Van Nieuwenhove of the WHO decide to leave for the border with Sudan, each by his own route, and to check whether there is a connection between the epidemics of Yambuku and the southern Sudan.

Peter Piot and Jean-François Ruppol fly to Kikwit. They rapidly establish that the prisoners do not have hemorrhagic fever at all, but liver inflammation.* A false alarm.

Monday, November 1, 1976
I have now been in Zaïre for one and a half weeks. According to the original plan of Stefaan Pattyn, my boss, I should be leaving in a few days to return home. The activities of the international commission have now settled into a working rhythm. Karl Johnson manages everything tightly yet pragmatically. The division of duties is clear, but much time and energy is lost on organizational challenges: maintaining contact with the field in Yambuku and Bumba, bureaucracy, transport questions, fuel... The epidemic has clearly weakened, but from virological, thera-

peutic, and epidemiological perspectives there are still mountains of work to get through. How is this new virus transmitted? How do we increase the chance of infected people surviving? Where did the virus come from? Does it come from an animal: a monkey, a bat, a rat? How can we prevent it happening again? Moreover, the outbreak has now well and truly caught the media's attention and has hence landed on the priority list of the international community, headed by ex-colonial power Belgium. Important folk have turned their heads toward our commission. We must succeed here.

In the evening after our day jobs the commission holds yet another meeting in the large Fométro hall. This time we are fourteen. Pierre Sureau isn't present, he is once again at Yambuku. Here in the tropics it has already been dark for hours when we start at 20:45. There are seven points on the agenda, followed by a list of countries that must help us with money, personnel, or logistic support. We agree to start future meetings much earlier, at 18:00, not to spend more time in long discussions that do not always deliver concrete results, and to regard the meetings, above all, as briefings, at which each constituent team presents a concise account of their activities. That can surely be done in an hour. If there are questions, they will be answered either immediately after or at the next day's meeting. For important topics of discussion the meeting will be interrupted at 19:00 for supper and resume after.

Under the agenda item concerning the supporting countries, Belgium is ranked second after the United States as the most important source of knowhow and dollars. The meeting decides to ask the Belgian administration for a Hercules that can shuttle between Brussels and Kinshasa every ten days, offering a regular air bridge to Europe. And then the commission requests "an extension of the missions of Drs. Peter Piot and Guido van der Groen of three to six months." I swallow hard. Peter left a pregnant wife in Belgium in the expectation that he would be home in a month at most. My Dina, also, will probably not rush to thank a husband who may well be absent for six months instead of two weeks, leaving her without a driver's licence, without full power over his bank account, and battling to hold down a full time job at the university while coping with two young children... After Belgium follow France, South Africa, Canada, and Zaïre itself. France and Canada also receive requests to grant their experts longer missions.

I am silent. Peter offers no more resistance than I. The stakes are too high, the "to do" list too long, the expectations cranked up too much, and the circumstances are not of such a nature that the other commission members would find my protest acceptable. Moreover, I find myself here in the bubble of a foreign land, a tense situation, new people, and every day another surprise. And then I haven't even been in Yambuku! Karl hasn't either, mind you. He is the undisputed leader here in Kinshasa, but in 1963, in Bolivia, he was immersed in the heat of the action when he discovered Machupo. It went so far that he became infected with the Bolivian fever himself and had to be evacuated in great haste to Panama. That had better not happen to me, but I do want to remain here longer than the two weeks my boss proposed. I sit, just like most people here at the table, between the anvil of my duties as a family man and the hammer of my scientific drive. The hammer wins, not for the last time in my career. Dina and my daughters will feel the hammer strokes harder than myself, and for this—the truth is the truth, however bitter—I have no other excuse than my own ambition.

There is yet another point on the agenda at this meeting that requires some reflection: the evacuation plan that we must call into play if a member of the commission runs a fever for longer than 48 hours.

Tuesday, November 2, 1976

Sukato undergoes his first plasmapheresis session. Half a liter of blood is tapped. The red blood cells are returned to him later. Some power of persuasion was needed, but the fact that he eventually only has to donate the "worthless, yellowish part of his blood that looks like urine" pushed him over the finish line. And also, of course, that he is paid, with sacks of rice, cans of sardines, or zaïres. However achieved, for the first time we now have antibodies from someone who contracted the dreaded disease and lived to talk about it.

We learn that in Belgium a special support fund is being established to fight the epidemic in and around Yambuku.

From Atlanta we receive the report that the new threadlike virus has also been isolated in the blood of the deceased Mayinga. It is the nameless virus that rules our lives. A few days later Pierre suggested: "Let us call it the Yambuku virus."

Wednesday, November 3, 1976

Sophie also undergoes plasmapheresis. Indeed, she wants to return to Yambuku as quickly as possible. To her, Kinshasa is as much a foreign world as it is to us. She has just been widowed, has a baby to care for, and has left behind people in Yambuku who miss her and who depend on her. She may return on the first flight back.

Meanwhile in Yambuku, Pierre Sureau has begun his 24th day in Zaïre. Just like the previous days, he is on the road with a translator and criss-crosses between the villages around Yambuku and Yalikonde. He leaves really early, at 05:00, because it is the rainy season and the punctual cloudbursts in the later afternoon often make the roads unnegotiable for his Land Rover. With marvelous precision Pierre records everything in a diary—distances, the names of the people he has interviewed, ages, family relatives, what people were doing the past days and weeks, where they had been, with whom they had been in contact, their symptoms, who in the family has died, and the numbers of the blood samples that he has taken—day after day, indeed sometimes fifteen pages per day.

A name. The commission agrees that it cannot just be "hemorrhagic fever virus" or "thread virus," because there are several of these. In the evening we stand in the passage at Fométro before a map that is not very detailed. Yambuku, for example, does not appear on the map, but not far from where the village must be, a river with the name "Ebola" flows. Perhaps that is better than "Yambuku"? Lassa virus and Marburg virus are named after the places where they first came to attention, but now the residents of those places must forever put up with this branding… Karl thinks "Ebola" is a good idea. His "Machupo" is also not named after the Bolivian village San Joaquin, but after the Rio Machupo that flows past it. Moreover, "Ebola" is easy to pronounce in all languages. Ebola it will be!

Thursday, November 4, 1976

Peter Piot, Karl Johnson, and Jean-François Ruppol ready themselves to leave for Yambuku.

I meet the kind Belgian couple Luc Van Damme and his energetic wife Frieda Behets. Luc is a veterinarian and has already worked for a few years in Kinshasa for the World Food Organisation. He can secure extra equipment for my lab.

I also get to know Bill Close better. The presidential doctor is also director of Mama Yemo, the biggest hospital in Kinshasa. He tells me how rotten the Zairean governmental administration is, in all departments. If you want something done, you must have *matabiche* in your hand: to set the administrators to work, to find out who has your file, to have your file extracted from the pile and opened in plain view. But I grasp that there is something wrong with Zairean society beyond this when Bill tells me that half the children in Mama Yemo have symptoms of undernourishment.

By contrast, Bill is well-disposed toward Ngwete Kikhela, State Secretary for Public Health in Zaïre. As a young buck, Ngwete managed to go to Canada where he washed dishes in order to earn enough to study medicine, and once he had returned to Zaïre he had sufficient drive to be promoted to his position today.

The CDC in Atlanta sends its research results to Kinshasa. The tests that they have undertaken in their gleaming BSL-4 lab on the sera that I studied in Kinshasa have yielded perfectly matched results. My lab may well be primitive, but I have worked in it with beautiful precision. Flems are not brought up to beat their own chests, but for the first time in my life I am proud of myself.

In Yambuku, Pierre concentrates today on his direct environment: the people around and in the mission post, where 38 people who stayed died from the disease. Among them are the Flemish sisters Beata, Myriam, Romana, and Edmonda; the hospital chaplin Father Germain Lootens; and six Zairean hospital staff members. Pierre interviews 11 families in Yambuku, who together mourn 16 relatives.

Friday, November 5, 1976

In Porton Down, England, in the Toxic Animals Wing of the Microbiological Research Establishment, scientist Geoffrey Platt is working with a small piece of liver tissue from a guinea pig. The laboratory animal has been killed by a virus from southern Sudan, where over the last summer an epidemic of hemorrhagic fever has raged. Maybe it is actually the same epidemic as that of Yambuku—this question is being investigated by Joe McCormick and Simon Van Nieuwenhove.

For almost a month Geoffrey has been investigating the virus. He is just about to inject another guinea pig with infected tissue as part of the so-called virulence test. This is a test to find out how well the virus preserves its disease-causing potential when test animals are infected by successive generations of the virus. You inject the virus into an animal, wait until it becomes ill, extract the virus, and inject it into another animal, and so on. Geoffrey is wearing protective clothing, using breathing apparatus, and working with the utmost care.

But suddenly, Geoffrey lets his guard down and relaxes for a fraction of a second, and a single muscle that is not one hundred percent under control jerks his hand and he pricks the top of his thumb. The injection needle pierces three layers of thin laboratory gloves. It occurs so quickly that Geoffrey questions if it has really happened. Rooted to the spot, he stares at his hand and then whirs into action. He pulls his gloves off, pinches his pricked thumb hard across its middle, and rushes to the neighboring room, where he dips his hand into disinfectant. Two minutes. Then he looks. No blood, no sign of a prick mark. If that means that no hole was made, he has been lucky. But if the hole was so small that it closed immediately after the prick, it could be very bad news. Hopefully the disinfectant penetrated the wound and did its work.

Geoffrey leaves the lab and reports the prick incident to the safety service of Porton Down. After a brief investigation he is sent home with a thermometer.

In the Fométro buildings in Kinshasa it becomes quieter by the day as yet more commission members leave for Yambuku. At present I remain behind, together with the American epidemiologist Joel Breman and the French hematologist Danny Courtois. While I continue my lab tests, Danny examines blood samples. It has been agreed that we too will depart for Yambuku on November 16 to construct a field laboratory with a secure isolator. I have just found four specimens with antibodies against Ebola. Sukato undergoes a second plasmapheresis session.

At last I have an opportunity to phone home. I confirm that there is still a world beyond Zaïre. How good it feels to hear the trusted voices once again.

Also good: I can now knock tennis balls on the head! Since turning twelve I have played tennis—not enough to be outstanding, but I am a strong C-player, a good amateur. I find it fantastic that I also have this valve to let off steam here.

In the afternoon Pierre Sureau is much uplifted by the arrival of Jean-François Ruppol, Peter Piot, and Karl Johnson in Yambuku. Karl knows what it is like to be isolated in the jungle and has brought a carton of American cigarettes and a flask of whiskey for his French colleague. Yes, we don't always set a good example simply because we do everything we can for the health of others! The three install themselves in the former house of the medical assistant, where Pierre is also lodging. They have much to talk about. Pierre learns that his own suggestion hasn't been chosen for the name of the new virus, but rather the name of a river. He can live with it, no problem. By the time that everyone has been brought up to date following dinner and goes to bed, Karl's flask is as good as empty.

Pierre Sureau has a restless night, with a headache, dizziness, diarrhea, and nausea. He must get up a couple of times to vomit.

Saturday, November 6, 1976
Pierre isn't feeling better and stays in bed. Karl Johnson replaces him during the daily rounds through the surrounding villages. Fortunately, Pierre doesn't develop a fever, and in the course of the day he perks up. It was probably the whiskey.

Monday, November 8, 1976
In England, Geoffrey Platt awaits his future after the needle stick injury. He has received instructions from Porton Down to report any rapid increase in temperature at once. Geoffrey and his wife Eileen can barely sleep at night due to worry but remain strong for their two young children.

My lab work is rapid but does deliver results. Among the 109 sera from people who have possibly had Ebola, I find 12 with antibodies against the virus. That means 12 new candidates for our plasmapheresis program.

In Yambuku, Pierre resumes his work. Of the 213 people who died between September 8 and October 21, according to the report of Sister Marcella, Pierre was able to document 148 in 21 different villages. Of the 148 deceased, 103 were the first to be infected in their family and then transmitted the deadly disease to at least 45 other family members. Three quarters of the "primarily infected people" were extremely likely infected in the hospital at Yambuku, probably by an injection with an unsterilized needle. Sureau had questioned roughly 300 family members of the deceased and had drawn blood from over 100 of them because they had at some stage had symptoms of the "Yambuku sickness." There were no roadblocks between the villages in the area that he had covered. Thus people still came to the hospital, although it had in the meantime been closed. Between Yambuku and Bumba there were indeed barriers to limit the epidemic.

Tuesday, November 9, 1976

The epidemic dies out, and for Pierre Sureau it is time to return to France after two visits to Yambuku. Père Dubois and Sisters Marcella, Genoveva, and Mariette give him a hearty farewell. They enjoy his company, and on his side, Pierre prizes their unstinting hospitality, help, and friendship. A helicopter ferries him to Bumba in 20 minutes. At about midday he boards the C-130, which is right on time. During his flight to Kinshasa, Pierre can sit in the cockpit, where his last view over the endless jungle and the Zaïre River is yet more impressive.

Upon his arrival at the Fométro building, Pierre gives me the blood samples that he has collected. During the daily commission meeting he presents his report. We also discuss the animals that could have transmitted the Ebola virus. According to the first findings, it looks like there is no connection with insects. Possibly the rodents or primates in the area play a role. We must ask the Zairean government to continue and widen the search for the animal host.

I give my own report. Among Pierre's samples I have already found ten Ebola-positive specimens. I also now have 800 milliliters of plasma from recovered patients. In order to be able to use it, we must be sure that there is no more living virus in the plasma, but because we do not have a BSL-4 lab, a portion of the plasma must be sent to the Center for Disease Control in Atlanta.

Wednesday, November 10, 1976

The team is again strengthened. Two epidemiologists have arrived from the CDC: Mike White and Stan Foster. Stan is a seasoned investigator who has crisscrossed much of Africa for the World Health Organization, on the trail of the smallpox virus. My initial encounters with Stan Foster are very good. It will be the beginning of a great friendship. On the radio we hear that Joe and Simon have arrived in southern Sudan, where they are now investigating a possible connection between the two outbreaks.

Joe and Simon think it is likely but have not found anything concrete. No definite link will turn up later, either. It will appear that the two outbreaks have nothing to do with each other.

The quarantine in the Ngaliema Hospital is lifted. The people still staying in the quarantine block may all go home. No one has fallen ill. Ebola has not taken Kinshasa in its grip.

A brief phone call with Kontich. I have my mother-in-law on the line. She describes how at that moment saplings are being planted in my garden, and I hear the voices of the children in the background.

In the early evening I develop a tic and suddenly feel wiped out from exhaustion. I make plans to visit the Museum of Fine Arts in Kinshasa. The garden there is a paradise of flowers, plants, and trees. Perhaps it will restore my senses? It strikes me today for the first time how extraordinary the light is at the equator. Nowhere else in the world will I see light like this.

Thursday, November 11, 1976

In the morning Geoffrey Platt reports to the safety service of Porton Down with fever. Panic! Not only for him, but also for everyone who has been in contact with him. Blood is drawn and studied under the electron microscope. The virus is definitely present. Under police escort Geoffrey is immediately taken to the hospital of Coppetts Wood in London. He is put in an isolation tent. The other 160 patients in the hospital are transferred elsewhere. Apart from the people who care for him—and they must not leave the hospital themselves—no one in Geoffrey's neighborhood is allowed to come, not even Eileen and his children. They are kept at home in quarantine. The research institute where the needle stick injury occurred is closed and all his colleagues are placed in house quarantine. A telex is sent to Kinshasa and the black news is reported to the national and international health organizations.

In the Ngaliema Hospital we had harvested our first bags of plasma with antibodies just days before the sorry news from England arrived. A colleague is infected with the hemorrhagic fever virus from southern Sudan. A hammer blow. England asks if we can send sera from recovered patients as quickly as possible. We respond immediately.

Although the Belgian Ministry of Foreign Affairs barely keeps the home front informed about the situation in Zaïre and how we are handling things, on this occasion Dina is told promptly. Thus she gets to hear that someone who does the same crazy work as her husband with the same newly discovered appalling virus has been infected and taken to a hypermodern hospital in London—although I'm sure the fact that this happened in a work environment literally and figuratively thousands of miles removed from that of her husband does nothing to calm her. At that moment I do not know that Dina knows.

The incident at Porton Down makes me think. Does it make sense to invest hundreds of thousands of dollars in a maximum security lab if just one wrong movement is enough to infect yourself? Why make an isolator with an internal wall as thick as a fist if you require staff to handle blood samples inside it with paper-thin gloves? At that precise moment I am working in Kinshasa in my field lab with an isolator, which has a wall that is just one layer of plastic, and it contains the virus just as well.

Friday, November 12, 1976

In the radio reports from Yambuku we now hear more often that there is a connection between clinically confirmed Ebola cases and the fact that the patients have taken part in burial rites involving deceased patients. Also by radio, Peter Piot lets us know that he is faring well, that no new Ebola cases have been noted, and that I must be sure to bring enough beer, soap, and medication against worms and mange if I come to Yambuku. As the team members there hear about what has happened in Porton Down, the news hits them like a bomb.

England notifies us that the sera we sent for Geoffrey Platt has arrived safely. They do not add whether Geoffrey has already been treated with the sera. The commission decides that team members who travel back to their homelands and who have been in contact with Ebola patients must remain in Zaïre for at least two weeks after their last contact. Pierre therefore remains in Kinshasa longer than planned.

Saturday, November 13, 1976

In London, Geoffrey Platt develops roughly the same symptoms as the victims in Sudan and Zaïre. He now has a 40°C (104°F) fever and there is blood in his stools and vomit. He is treated with all possible means: with the antibodies from Sophie's blood that we have sent but also with the newly isolated human interferon, an important chemical for the immune system.*

Geoffrey's hair falls out. He is completely weakened, extremely nauseous, and incontinent. His mental condition also deteriorates badly. He can no longer read, cannot remember that he ever could read, and asks why in God's name he is lying in this plastic tent.

In another three days I depart for Yambuku. In the meantime I start packing all the equipment and dismantle my lab. It is much more efficient to diagnose new patients on site and, cost what it may, we want to find more recovered patients and perform plasmapheresis.

Monday, November 15, 1976

Tomorrow I fly to Yambuku. But first we must digest a moment of horror when we learn that the Alouette helicopter that would have brought Peter Piot from Yambuku to Bumba for a meeting with the American Ambassador has crashed. The home front receives the message that a helicopter has crashed, but they are not told who was on board. Happily we in Kinshasa quickly learn that Peter was not on board. He doesn't like flying, a storm hung in the sky, and the helicopter pilots were drunk. Moreover, he had no interest in a trivial duty such as meeting a diplomat, although Karl Johnson had strongly urged it. Fuck you, Peter must have thought, and that would have saved his life. He is still in Yambuku, alive and kicking, but without a helicopter.

Pierre Sureau's quarantine has finished and he departs for Europe. I have gotten to know Pierre as a warm personality and a pure-blooded field epidemiologist who would rather be in the jungle than sitting in his office. Yet he has recorded his findings with dumbfounding precision in a diary. Pierre also has an eye and respect for the work of others. I will never forget his praise for my lab and the first results that I notched up. We will miss him.

That evening there is a farewell drink with shots of whiskey. At the gathering I speak with Bill Close. He has obtained a video camera from the American embassy, the latest new Sony model, with a little foldaway TV screen. Bill doesn't do much with it, but I am interested. There is no handbook, but by trial and error I get the thing running. It works much better than the small film camera that I brought from Belgium. Bill agrees that from now on I should keep the camera and take it with me to Yambuku. There I will often use it, and some of the images will travel around the globe.

At 23:00 we wave goodbye to Pierre Sureau as he boards a flight to Switzerland. In 1977 Pierre will receive a letter from Sisters Marcella, Genoveva, and Mariette, to thank him for the help he came to offer in Yambuku "at the moment others did not dare." I will see him later at a couple of conferences. Sadly, this great gentleman will die far too early in 1994, at the age of 68.

Tuesday, November 16, 1976

Yambuku at last! Stan Foster is the only team member to stay behind in Kinshasa. The aircraft carries us to Bumba, where we fully load two Land Rovers with our equipment. It is a drive of 150 kilometers. I sit wedged into the left back seat, in a horribly uncomfortable position, with a wall of boxes on my right, an unknown object that pokes me in the neck with the rhythm of the bumpy road, and on my lap the big box with the UV microscope. Through the little window conspicuously smart villages jolt past. Women sweep with brooms. Men sit in large reclining chairs. Shouting children accompany us momentarily. We drive for four hours.

Yambuku looks desolate, but we too receive the warm reception from the sisters about which Peter and Pierre were so lyrical. When we have unpacked everything, I settle into part of the convent right next to the place where we eat together. My instruction is clear and urgent: build a lab, install the new generator as soon as it arrives, and organize a space in which we can carry out plasmapheresis. Throughout my work I use my new camera. When I view my first images, Yambuku looks even more disconsolate.

Saturday, November 20, 1976

For plasmapheresis you need a centrifuge that not only spins the vials of blood around but also keeps them cool, otherwise too many of the red blood cells that should be returned to the donor die. Freezing point is not required, but to obtain the necessary 5 to 6°C in the tropics is difficult enough. Quicker than expected I have everything assembled. Male nurse Sukato is once again the first in line for plasmapheresis, this time in the middle of the jungle. We are 1,200 kilometers from Kinshasa and I suspect that we are scoring a first. Sukato, also, for that matter. With fifteen donations he will graduate as the world champion of plasmapheresis. If there should be more cases of Ebola, we can now immediately administer antibodies.

I have a talk with Sister Mariette. Her case intrigues me. Day after day the sister cared for her fellow sisters who are now dead. She washed them, removed sanitary towels, and had bloody vomit spattered on her legs and feet. Yet she did not fall ill. Is the virus less infectious than we think? Mariette states that she washed her hands frequently and faithfully followed the advice of Father Germain. He made her swear not to touch her eyes and nose with her hands. The father himself did not survive, but his words remained hanging like a mantra. Luck or discipline? The latter, I think. Also in the commission no one has yet been contaminated. In any case I am consequently careful not to rub my eyes or pick my nose. The other investigators follow suit.

In the evening it strikes me that there are many bats flying around Yambuku. My boss Stefaan Pattyn has already suggested that these small creatures could well be the host of the virus. Our colleagues back from southern Sudan also saw many bats hanging from the ceilings of spaces where infected people stayed. I must keep an eye on this.

From England we learn that Geoffrey Platt is now faring better. It is not certain that he has our serum to thank for this, because monkeys that were infected with the same virus and received the same treatment nonetheless died. When years later I meet Geoffrey face to face, we greet each other with a bear hug, like Russian friends. "If I follow my heart," he confides in me, "you saved my life, Guido, but my reason doesn't accept this hypothesis, because I have my doubts about the efficacy of your serum."

Sunday, November 21, 1976

Sunday, YBS day! No tennis in the jungle, thus the only hobby that remains for me in Yambuku is filming. I make outdoor recordings and stage the very first broadcast of the multinational TV station, which we baptize the Yambuku Broadcasting System. Naturally there is more to this game than my fascination with cameras and broadcasting. The members of the international commission hunker down at different locations within the mission post. I sleep close to the Flemish sisters and as a Dutch speaker I am their natural link to the rest of the international team. Peter is also a Flem. As an epidemiologist he concentrates on continuing the work of Pierre Sureau and is very often on the village roads. Indeed, I also travel, sometimes to film, but I spend more time at the mission post because I receive blood samples that I must subsequently analyze in our lab.

Thus we all have our daily jobs. Mostly we are content with this, so content that we sometimes lose sight of the astonishing drama that played out here. The scars haven't even faded by half. The residents of Yambuku bear them, because each has lost at least one dearly beloved, but the sisters are also fighting trauma. What does it do to you, to see people with whom you have lived for years die by the dozen, following a gruesome disease process? People who had invested all their hope in you, whom you had wished to save by any possible means, and who, despite this, slipped through your fingers? To subsequently see your fellow sisters pass away, on a ramshackle iron bed in a tropical hospital 6,000 kilometers from the Campine villages where they played hopscotch and grew into idealistic young women, and where they on some hopeful day took leave of proud yet sorrowful parents? Not to speak of the loneliness and isolation during quarantine, followed by the stress of arriving and departing researchers, men with accents from around the globe who are, of course, welcome, but to whom you must offer a sturdy stove, good meals, and decent shelter.

And so the idea was born to find some diversion, something in which everyone—local residents, aid workers, clerics, and commission members—can lose themselves. I team up with Sister Genoveva, who is always ready for a joke and whom I have come to call "Geno." Several days ago we sketched a plan to film interesting activities in and around the mission, to create a daily "TV-program" for the Yambuku Broadcasting System. The waggish Geno becomes my TV reporter. I am cameraman, journalist, producer, and director. We quickly agree on the tone of YBS broadcasts: humor and upliftment! For sound we have the pick-up and LP's of the mission. The mellifluous sounds of George Zhamfir's panpipes win our votes for the signature tune.

I have already shot a lot of footage at the mission, but Geno and I have also visited the villages where the sister tries to improve the living circumstances of the residents. She explains how to make bricks from red clay to build stronger houses. Geno also wants to help lift the people from the ground, literally, through the provision of tables and chairs. I film and admire.

In the late afternoon my show is ready and I plug the video camera into a small TV set that I have positioned on the veranda. At 18:00 dusk falls and the jungle sounds swell, a delicious moment each day that at once creates an atmosphere of intimacy and expectation. It becomes dark in less than an hour, during which my audience trickles in. Everyone sits near the edge of the large veranda around the little screen. The broadcast begins with an introduction by Geno. Film footage alternates with cartoons that I have drawn, adverts such as "Blood with antibodies—10 Zaïre, call Papa Machupo," and a sheet of paper on which I have written "Interlude" with stars around its border, accompanied by a soundtrack of arousing choral music. There is laughter and much commentary. Some local residents sit and stare at the box open-mouthed.

One of the stories that I filmed for YBS is about our new generator. The mission's was not powerful enough for the international team's equipment and thus we had organized another one. This machine had to come from South Africa, be flown to Bumba, and driven to Yambuku on the back of an open truck. The whole village descended on the mission for the spectacle. With great jubilation we saw the truck arrive in the distance, dust clouds swirling upward. The truck parked neatly in front of the mission. So far, so good, but now the problem was to shift the leaden monster to its final position, without a crane or any other lifting equipment. Fortunately the African countryside has an inexhaustible supply of one resource: manpower! After noisy debate, it was decided to make a ramp of stout wooden planks between the truck's deck and the ground, down which the generator would be allowed to slide, controlled by muscles. Not mine of course, because I had to film how all the flexing biceps and sweating backs were braced against so much gravitational force, and how pensively the sisters stood watching from a safe distance. Inch by inch the danger edged lower. As soon as it stood on the ground, cries of joy resounded. Round poles over which we could roll the machine to the entrance of the building lay ready. Projecting parts of the beast were quickly removed, as it was much too big to fit through a door or window. Eventually we had maneuvered the generator to its station, safe and sound. The local populace could congratulate itself, the sisters made the sign of the cross, and I had captured a fantastic YBS story.

A lot of soccer was played in Yambuku and the surrounding areas, and that also yielded material for YBS. An extra large crowd attended the

Yambuku-Yandongi derby: cake on a plate for reporter van der Groen. It was not immediately clear where the boundaries of the playing field lay—indeed, that would be determined by the linesmen—and there was also no perimeter. The spectators stood more or less at the side of what a reasonable soccer field would be. When a goal was scored, everyone stormed onto the field and for minutes there would be dancing and singing, until the referee felt it was enough and, with a sharp blast on his whistle, brought the celebration to an end. Just until the next goal, mind you. Never had I experienced so much pleasure from soccer. In the evening the supporters of both camps were invited onto the veranda to see my match report. A good sixty or seventy people packed in around the tiny TV set. For most of the audience, it was the first time in their lives that they had seen moving pictures. It conferred on me the status of village magician. In my turn I found it fascinating to be the witness of the comical existential experience. At each goal there was again jubilation as if it had fallen from the sky entirely unexpectedly.

Wednesday, November 24, 1976
I have not even been here ten days, but it is stunning how a person adapts to situations that not long ago seemed utterly alien. In the blink of an eye you develop new routines, which feel as if you have never done anything else in your life. The epidemic tails off further, happily, but every day I receive yet more samples from people who were possibly sick, and which I analyze in my lab. Moreover we must be prepared for the risk that the outbreak still smolders here or elsewhere and at some moment will flare up.

I am reinforced by Del Conn, an American volunteer of the Peace Corps who doesn't have anything on the go right now. I could well use assistance in my lab and thus I take Del under my wing and teach him some laboratory techniques. He appears to be a first class pupil and good company.

Sunday, November 28, 1976
After yet another week working in the lab and field, we accompany members of the international commission and some mission sisters on a day outing to the Dua River, which together with the Ebola flows into the Mongala and thence into the Zaïre. In groups of two we scramble aboard dugouts. At the bow and stern of our hollowed-out tree trunks, paddlers

stand prepared to launch us out over the jet black water. I take my place behind my French colleague Danny Courtois, my 16 mm camera at the ready. I am astonished when I note that the sisters in their crisp white outfits drop their customary seriousness and giggle like girls at the sight of Danny. This fervent lepidopterist holds a small fishing net in one hand, which he flicks out at each butterfly that flutters past, and in his other hand a little jar to contain his haul. Everyone has a good time. In a riverside village we eat roasted larvae and I again see the familiar pattern: men talking to each other from their deck chairs and women who are stamping manioc, busy with a broom, or balancing massive bunches of bananas or jerry cans on their heads while walking gracefully, their hands free. I contort myself into all manner of positions in order to shoot the best shots.

The return journey is much quicker. Our paddlers have drunk palm wine and throw caution to the wind. The canoe begins to roll dangerously. After an interval I note that each time the boat rocks we ship some water. We are diving steadily deeper into the Dua and falling behind the other dugouts. We are sinking. With outstretched arms Courtois tries to keep his little net and jar dry. I do the same with my camera and consider myself lucky to have had the foresight to leave the expensive Sony behind at Yambuku. We have now sunk so deep that we have just about lost sight of the other boats. Because we too have imbibed the palm wine, it takes a long time for us to realize that there is just one thing to do: push the pirogue to the bank and bale it out with big leaves.

Once that is done we again venture out against the painted backdrop of a tropical sunset, and this time somewhat more carefully. We laugh and sing. I thump out the rhythm on the wooden side of the little boat until we are drowned out by the sounds of the jungle. They seem to come from an amplifier. We fall silent. The black water reflects the uncannily clear starry heaven that drifts over our heads. Every cameraman will say it: the most beautiful things in life cannot be captured on film.

Monday, November 29, 1976

My assistant, Del Conn, is not at his post. He has a high fever. Of course everyone thinks of Ebola, but not I. I have monitored Del for the past week in my lab and saw that he took no risks. But a decision is a decision, and for team members with a high fever, evacuation to South Africa is compulsory. Naturally I would be able to test his blood myself, but it can take days before antibodies can be detected and, according to the agreed procedure, we do not wish to lose any time. I film how Del, weak and slow, lies down on the mattress placed in the back of Karl Johnson's Land Rover. In Kinshasa he will be put aboard a military aircraft, on a firmly secured bed in a plastic isolation tent, to be flown to Johannesburg. But first there is the bumpy road to Bumba. Del is dreading it. Everyone tries to raise his spirits.

Tuesday, November 30, 1976
Bad luck for Del. Due to bad weather his flight is rerouted via Angola, so that he is underway for extra hours.

Wednesday, December 1, 1976
I have just had an exceptionally bad night, but my morning temperature is only 36.5°C (97.7°F). After a savory meal with pickled herring and a short siesta I recover during the course of the day. We learn that Del has arrived safely in Johannesburg and his condition is stable. So you see, I think. The man will later recover fully, although it would never be clear exactly what it was that he had. In the evening I measure my temperature anew and it is once again 36.5°C. No fever, no Ebola. Diagnosis: a whiskey overdose.

Thursday, December 2, 1976
Karl Johnson returns sick to Yambuku after staying several days in Bumba. He is visibly tired and complains of diarrhea.

Just one more week and our time in Yambuku will be up...

Friday, December 3, 1976
Stefaan Pattyn visits us. Peter Piot and I bring our chief up to date with an extensive explanation of what we have achieved in the field and in the lab. I also tell Pattyn about the video recordings. He suggests that my images can serve to seek sponsorship in Belgium for the mission's sick-bay.

Pattyn has another surprise for me: an audio cassette upon which Dina, my daughters, and brother-in-law sing happy birthday to me. The rest of the tape is filled with reports about family life in faraway Belgium.

Karl is better again.

Saturday, December 11, 1976

It is my birthday and in my notebook I write: "34 years old. I see no use in expanding upon this."

High level visit today: State Secretary for Public Health Ngwete comes to put new heart into our team and the local population. Just in time, because soon we will be gone. During the past few days we have readied our notes, research samples, and other material for transport to Kinshasa. For the sisters, our departure closes what has clearly been the most trying period of their lives. But they are not to be left to their fate. David Heymann, a 30-year-old doctor from the CDC, is coming to relieve us. If the outbreak should flare up again, there will be immediate intervention under his supervision. Thus David will not be spending New Year's Eve snug at home in the United States. Full credit to Heymann, because with his help we can take our leave of Yambuku with our hearts a little less heavy.

We are now on our way to Bumba. How I swore over the past few weeks when we were on the road in a Land Rover. The sturdy leafsprings were punished to the utmost and each hole catapulted me from my seat for the umpteenth time. But now I already anticipate feeling nostalgic for these roads, partly because I have in the interim learned why they are so bad. The population wants to keep them that way. The people are dead scared that motorized thugs will all too easily reach their villages, to rob and to rape. The holes in the roads discourage them. And if they don't prevent the scoundrels arriving, then at least they will hinder their advance to a fair degree. That will give the tom-toms enough time to announce the approaching evil, so that the people can take precautions. Tom-toms are a centuries-old yet surprisingly quick and efficient means of communication in this remote corner of Zaïre, where radio links hardly exist, let alone telephones.

The trip to Bumba must have been too easy, because during the flight to the capital we had to endure a tropical storm. I reflect upon the day that I arrived in this land, and on the happenings in the Ngaliema Hospital and in Yambuku. We shall nonetheless remain in Africa for a couple of days yet to write our reports. I tease Heymann, because the prospect of seeing Dina and the children again just in time for the Festive Season is heartwarming. But the lump in my throat says that I also envy Heymann. I try to sleep, without success. The storm subsides and somewhat later we land in a drenched Kinshasa that appears a little grayer.

Wednesday, December 15, 1976

Since returning to Kinshasa, we fill our days with writing reports. I convince a few tennis players to hit a ball with me. Although we wait until evening falls, it is still 29°C (84.2°F). After the little tennis party, Peter Piot, Simon van Nieuwenhove, Luc van Damme and his wife Frieda Behets, and I decide to eat cossa-cossa one more time. These delicious river prawns, so large that they can pass for lobster tails, are a specialty of Kinshasa.

We hear that somewhere in the world a DC-10 has lost an engine during flight—this is the aircraft that must fly us to Belgium. All DC-10s are grounded for a few days.

Thursday, December 16, 1976

In the luxury hotel Memling, where there are always many Belgian guests, we make the acquaintance of the Sabena captain who will be flying us to Brussels tomorrow. When he hears that we have already been in Zaïre for two months and haven't tasted the nightlife of Kinshasa even once, he finds it to be his civic duty to introduce us to this pleasure. It is 2 a.m. when we accompany the captain to his hotel after an enjoyable evening. We are sure that our flight, with a planned departure time of 5:45 a.m. from the N'Djili airport, will be postponed for several hours.

Friday, December 17, 1976

Less than four hours later the captain is there waiting for us, freshly shaven and wide awake, standing beside his DC-10. He proposes that we sit in the cockpit. Peter turns this down, but the unrealized pilot in me cannot reject this offer. I shake the hands of the co-pilot and flight engineer. The captain is immediately concentrating on his job, because the traffic control tower announces that another flight now enjoys priority, and thus the flight details must be changed. Then comes the green light, and we may take off. I keep my eyes on the captain. He is still sitting nicely with his head up, but once we are above the tree tops, his body sags and his head falls to one side. I give the co-pilot a surreptitious wink. "Mother, that took a long time," he says. The man appears to know about last night's escapade and had agreed with the captain that he would take over as soon as we were well and truly in the air.

After a short stop in Athens we land safely at Zaventem at 18:45. The reunion with my family and relatives is warm, but also feels a bit strange. With the long beard that I have grown in the interval, my children think I look like a missionary. My sojourn in the tropics has left its imprints.

Monday, December 20, 1976–end of January 1977

We spend the first workdays after our return writing out the information that we have gathered. The international team conducted literally thousands of interviews in and around Yambuku, and these must be processed. Some patients were questioned by several team members, often yielding striking discrepancies. To put all the answers in the right order and report them correctly is a titanic task.

When we combine the epidemiological information with the laboratory data, we conclude that in Zaïre, in the period between September 1 and October 30, 1976, a total of 318 people were infected with Ebola and 280 of them, or 88 percent, died. New infections peaked on September 23 and 24. When Pierre Sureau and Peter Piot arrived in Yambuku in the middle of October, the epidemic was past its high point. Thus the course of the outbreak was not curbed by the international team, but rather by the decrees of the village chiefs who saw to it that no more people went to the hospital, that the sick stayed in their huts, and that they had as little contact as possible with people who were not infected.

An estimated 30 percent of victims became infected when they received an injection with a needle or syringe that was not sterile, probably in the Yambuku mission hospital. With the long queues and the express desire of patients to receive an injection of any kind, the sisters were placed under high pressure. Moreover we know that hypodermic needles were stolen from the hospital and that some people, often for payment, gave injections to their fellows in unhygienic and, of course, non-sterile conditions. That, too, contributed to the spread of the virus, just like the traditional practice of making cuts in the skin and rubbing a plant extract into them. However, the greatest number of infections occurred through contact with the body fluids of a patient: blood, vomit, and stools. Such contact took place particularly during funeral rites in which relatives washed the deceased, or—in the first phase of the epidemic—due to the fact that carers did not wear aprons, masks, or gloves.

On a dark day in January, Peter Piot and I visit the abbey of the sisters of the Holy Heart of Maria. We had already spent much time at their mission and must address several issues that we had identified there. We do not intend to backbite, because in Zaïre we had already indicated to the sisters Genoveva and Marcella that it was likely that the organization of work in the hospital had contributed to the spread of the virus. In the meantime these suspicions had been confirmed. We want lessons to be drawn from the epidemic and it is here, in 's Gravenwezel, that it can be determined how the Holy Heart of Maria can best achieve that. It is wonderful to help people, so long as it happens in a constructive manner and with a basic infrastructure that bears up in the long term.

What happened in the Yambuku Mission was instigated by a philanthropic but short-sighted impulse. You set to work as a cleric in an area where almost no one can read, write, or count. Hence you start with teaching. You don't do that for one year, but for decades. And that goes outstandingly well. It is just so sad that many children never or only irregularly find themselves sitting on the school benches. They are sick or must help at home because a relative is ill. And thus you now also wish to address health care. But then it goes wrong. Mother Superior tolls the bell of the Fathers because they are good at accessing financial resources. A miraculous sum is collected, with which a small pharmacy and an operating theatre are built. Magnificent, mind you, because such a facility exists nowhere else within a radius of 100 kilometers. But no one asks how long you are going to do this, how you must organize everything, what budget is required to recruit well trained personnel, or how you are going to solve the hundred-and-one practical problems that come with such an initiative. In short, the sisters of that early era had begun a megaproject, the scope and consequences of which they never guessed at.

Nevertheless there are examples that show it can be done better. The American missionaries conscientiously provide for at least a five-year budget before starting a project. They fly in their material and food needs instead of transporting it by boat or in trucks, because in the latter case half the cargo disappears regularly. And the personnel of an American mission are rotated every two years to another post, to prevent too much money sticking to the fingers of local officials.

Armed with these arguments and considerations, we travel to the North Campine and talk with the sisters there, in a great gray, silent, cold abbey. The contrast with the village of Yambuku we have left behind could not be greater. But our visit delivers nothing. We say what we have to say and the sisters do the same—they stress above all else the need for money to attract a doctor to the mission—but our words clearly lack carrying power. And I, the heretic of Kontich, do not get my heart to accept the need to convince these faithful souls with more forceful language.

In the beginning, the sisters of Yambuku worked with the best of intentions and astounding dedication, even scorning death—I had witnessed that for myself on several occasions in Zaïre. I know very well that the path to Hell is paved with good intentions, but in this regard it was difficult for me to scold the sisters. Secondly, you can throw a stone—perhaps even a whole truckload of rocks—in the direction of the Zairean administration. The government never included the aid of the sisters structurally in a project and never questioned the financial resources essential for an initiative that concerned public health. Thirdly, it cannot be excluded that the sisters—by helping long rows of patients daily at such a tempo through all the years, however imperfect that help may have been—had, in fact, saved many more lives than if they had instead treated only ten or twenty people per day, due to sterilizing the needles between patients. In that case they would have had to send the remainder of the long queue away, and eventually the people would have stopped coming and health care would have been limited to the village healers. Two things are clear: the epidemic should never have happened, but the fact that it did is due to more factors than the well-intentioned amateurism of Flemish nuns.

Early 1977

In Kinshasa I worked in a lab with an isolation tent made of flexible plastic. Inside such an isolator the air pressure is lower than outside. Hence no air can escape through a puncture—and thus also no bacteria or viruses. The air is continuously sucked to the outside through a so-called absolute or hepa-filter, which traps even the smallest particles. But a system or procedure is never stronger than its weakest link, and Geoffrey Platt's needle-stick injury in Porton Down's maximum security lab hit me between the eyes, the arbitrariness of "maximum security" becoming starkly apparent.

Stefaan Pattyn wishes us to undertake further research on Ebola at the Institute for Tropical Medicine. Okay, but then we must build an MSL. Problem: the cost of an MSL secure enough for that type of research, such as that at the Center for Disease Control in Atlanta, runs into the millions. We do not have that kind of budget in Antwerp. I propose to my boss that we lower the bar somewhat and create a better version of the isolator I had contrived in Kinshasa inside our existing laboratory. The plan is that I first survey the territory in the United States and learn how you handle Ebola in a state-of-the-art MSL. From this I can derive the minimal requirements for us to work safely and professionally at the ITM.

Since Yambuku I have had a first-class contact in Karl Johnson, head of the BSL-4 lab at the CDC. We correspond regularly, and Karl invites me for three months of training in Atlanta. I can lodge with Stan Foster. I got to know him too, in Zaïre. My visit will fall in the summer months, such that Dina can come over for a few weeks with Ilse and Sonja.

Monday, May 30, 1977

Joel Breman, the epidemiologist with whom I worked in Zaïre, fetches me at the airport in Atlanta and gives me a ride to Foster's place. This will be my first long stay in the United States, where everything is bigger and in greater abundance. My first glimpse of this is on the twelve-lane road that encircles Atlanta, an expansive, green city with a few skyscrapers in a relatively small center, surrounded by countless residential suburbs. Joel drives me through such a suburb with neat, fully detached homes with trim front yards and robust pick-ups in front of the doors. We are in the south of the US but I do not see many black people here. They live in other areas.

Stan Foster and his wife, Dotty, a Guatemalan teacher, continually receive lodgers from all continents. They thought at once it would be great to have me. As devout Protestants, they carry high the flag of love for one's fellow man. Their hospitality was so much taken for granted that it touched me deeply. Stan and Dotty show me their house and make me cross my heart that I will from now on call this place "home." The couple's teenage children find this umpteenth guest to be the most natural thing in the world. One of their four children, a son, is attending a scout camp.

June 1977

In the hospital of Tandala in Northwest Zaïre, a nine-year-old girl dies of hemorrhagic fever. Ebola is later discovered in her blood. The girl is the first case since the epidemic in Yambuku and southern Sudan one year earlier. Investigation in the region around Tandala shows that in 1972, two people had already died of Ebola, and that about seven percent of the residents have antibodies against Ebola. After the nine-year-old girl, a few other people contract the disease, but none of them die.

Stan is a smallpox specialist and spends much of his time in foreign countries, often on the orders of the WHO or the CDC. Now and then Dotty accompanies him. Also now, they must be away for a while. But that isn't a problem, I do the honors: show new guests around and take the children to the sports club. I use David Heymann's car for this. He was the man who took the baton from us in Zaïre. He had now returned to Atlanta, but had, in the meantime, already left on another mission and doesn't need his car. Dave's heart is as big as that of the Foster's. The only condition is that I drive the car to his parents in the neighboring state of North Carolina at the end of my stay. That seems a fair deal to me, and so the vehicle for my daily commute is arranged. The CDC is housed in a towerblock that is almost entirely jacketed in glass that reflects the clouds. The colonial architecture of the "tropical institute" has its charms, for sure, but this is something else.

The reunion with Karl is congenial. "Hi, Guido, there you are, finally!" he calls to me. Regardless of whether you are professor, director, team leader, or technician, everyone here addresses you by your first name, and the hi's and hey's bounce back and forth. Karl immediately sends me to Frederick Murphy, the world famous virologist. "Fred sits next door, go and see him." Okay, so it's Fred, then. With him I will take many electron microscope photographs of viruses, one of his specialties. The Americans invest heavily in their scientific developments, that much is clear. The laboratories are top-notch.

In the CDC they have two types of MSLs. Inside one type of maximum security laboratory a stand-alone isolator operates under negative pressure, and the researcher undertakes all manipulations within it behind a thick glass wall by placing his arms in long plastic gloves sealed to holes in the wall. There is also negative pressure in the space around the isolator. The air from the isolator and the lab itself is cleansed by at least two absolute filters and thereafter recirculated. In the second type of lab the researcher dons a plastic suit that has its own air supply. Thus he is protected from the virus, which in principle can drift freely in the lab. The advantage of the second type is that your whole body can move more freely to carry out manipulations. Hence you move through the lab space in your own isolator, with a wide hose attached to your suit through which clean air passes for breathing.

I also make my acquaintance with Jim Lange. He will be my training supervisor and introduce me to the dodges and procedures of handling viruses in a BSL4 laboratory.

That evening, at the dinner table of the Fosters, I read the newspaper. I scarcely find news from outside America, let alone Belgium. In this regard the United States is a hamburger: on its east and west coasts people still want to know something about the outside world, but between the halves of the burger roll the meat—the states in the middle—is only smothered in its own gravy. However, it was easy for me to speak as the resident of a small kingdom much more exposed to foreign countries than the US. I shove the newspaper aside and write a letter to Dina and the children about preparations for their visit. Long distance calls are terribly expensive and the Fosters, with their incredible hospitality, would forbid me to repay that part of their telephone bill.

Several days later there are local elections in Georgia and I am granted leave. Let's have a little holiday weekend, I think. Thus I go to the supermarket to buy some beer, but that is beyond the mentality of this state. The cashier says: "Sorry, sir, but we cannot allow you to take this amount of alcohol with you, because of Election Day." On Election Day, no alcohol may be sold in Georgia, an old regulation to prevent votes being bought with free alcohol, which will only be repealed in the year 2000. I throw back the comment that I come from Belgium and don't even have the right to vote here. Belgium—now that she must ponder. The heart of Europe? "Ah, Brussels," she says full of enthusiasm. Yes, I state further, and at home we drink liberally on election days, to give us the inspiration required to know for whom we should vote. The lady has no answer to so much European logic and calls the manager. He dismisses my arguments with laughter and makes it clear to me that he cannot allow this sale to proceed. We conclude our exchange with a little contract. At 22:00, just after closing time, I return to the supermarket. Just as agreed, the friendly manager stands at the door of the supermarket ready with my chosen cargo that he also helps me to load.

Tuesday, June 14, 1977

The Fosters' house is an international reception center. Today I receive a family with three children. The parents take a good taste of the beer that I still have left over from the election weekend.

After two weeks at the CDC I feel completely at ease. The work in the MSL is going very well, but it is also a little lesson in constant stress. If you forget to bring something into the lab, you cannot just fetch it but must go through the whole procedure again. Into the inner airlock, strip off all protective gear, into the outer airlock, don normal clothes, fetch the missing stuff, back into the outer airlock, strip off normal clothes, enter the second airlock, and don protective garb once again. A good school indeed for my looming future work in Antwerp. For each task I wish to perform, I write down the steps that I must follow, what materials I need for the task, and how much. This is a monk's work, but it will save me a great deal of swearing.

That evening I write in my notebook that I miss my family and will be happy when they arrive soon.

Saturday, June 18, 1977

It is pleasant to make the acquaintance of the sixteen-year-old Paul Foster, who is back from his scout camp and is received by a complete stranger in his own home from an even stranger country—and finds it agreeable.

A letter arrives from Belgium, which has been six days in the post. The children have done excellently at the music school. I glow.

Monday, June 20, 1977

At the CDC I assess what needs to be assessed. Training supervisor Jim stands by me in word and deed. I go over the working of filters, what happens with waste... everything goes into my little notebook, in which there is also a comment that the gas in America is really cheap.

Karl Johnson gives me a copy of a videotape in which he has collated all my Yambuku reports. During the midday break I look at them, together with Patricia Webb, Karl's wife. Because Patricia was not in Yambuku and I have apparently forgotten much, it is exciting to see the images again. I do not know any longer what is coming up, and had no idea whatsoever that at a given instant I had filmed Leslie, Karl's secretary. He had allowed his eyes to fall on her. That also appears in the footage. There is an uncomfortable silence. Patricia rocks on her chair.

I decided to phone Dina once. They arrive on July 10. I must work during the first days, but thereafter we can all go on holiday and reconnoiter the land.

Saturday, June 25, 1977

Patricia and Karl have invited me over. This globetrotting research couple lives a good 40 kilometers from Atlanta in a sea of green and peace. Their attractive house has big sliding doors. In the spacious living room I linger by a great display case in which Patricia has assembled model giraffes that she has collected from all the countries she has ever been in. I compliment the Johnsons on their dwelling. One disadvantage: this nice resort lies in Ku Klux Klan territory—a fact that came to their knowledge after they had already settled here. Patricia tells me that men with pointed hoods sometimes gather in the nearby bushes for their rituals. I imagine how the Johnsons behind their sliding doors stand and watch the glow of burning crosses.

Friday, July 1, 1977

I can still use Dave Heymann's car, because he is leaving for Cameroon. Together with colleagues at the Institut Pasteur, he will check whether the pygmies of that country have antibodies against Ebola. Patricia Webb also departs. She is going to Sierra Leone.

Saturday, July 2, 1977

I visit the city center. It strikes me how colorfully the passers-by are dressed. Peachtree Street, the Carnaby Street of Atlanta, shows me the broad spectrum of American fashion, from women with barely covered busts to well-to-do blacks in tailored suits and white shoes. I enter the lobby of a gigantic hotel and my jaw drops at the variety of shops and entertainments, such as video games. Two hotel guests are playing at dueling cowboys. As they shoot, "Bang!" flashes on the screen or "I'm dead!" and a tombstone. Crazy! People wait in a queue in order to buy photos of themselves taken by a video camera and printed by a computer. Others shuffle on to the glass lift, which hangs on the outside of the tower block.

Sunday, July 10, 1977

As I drive in the late afternoon to fetch Dina and the girls, the Fosters' pick-up decides to play up. The return trip will have to be by taxi, the final discomfort of what has evidently been a troubled journey. It began yesterday morning when my father, who was driving Dina and the children from Kontich to Luxemburg, took the wrong exit from the highway and drove a stretch in the wrong direction. Because the flight path passed over the Arctic Circle, from Luxemburg the aircraft had to stopover at Reykjavik. From there it flew to New York, but the air-conditioning of the aircraft didn't work and some of the leaking coolant fell onto Dina's clothes, so that she arrived in America soaking wet. At the airport in New York everyone from Iceland was treated as a drug smuggler; the security checks lasted four hours, so that Dina and my daughters missed their connection to Atlanta. That was 4 a.m. this morning. Dina hadn't managed to get a wink of sleep because she had been keeping an eye on six- and seven-year-old girls. Happily, Sonja and Ilse had behaved in an exemplary fashion, while mother had to secure another flight to Atlanta. When that had been arranged and Dina clambered into the pick-up after 30 hours of traveling, a big oil leak made it plain that it had conked out. All baggage out of the trunk, wait for a taxi, all baggage into the taxi's trunk... Welcome to America! From now on it can only get better.

Monday, July 11, 1977

While my family adapts to the heat, in the CDC I discover the isolator in which astronauts Armstrong, Aldrin, and Collins were briefly housed after their trip to the moon. The CDC had bought it for one dollar. Last year the isolator stood ready during the Zairean epidemic to evacuate infected American commission members to the US, but it wasn't needed.

In the evening we go to an American football match. We leave the car behind at one of the extensive parking areas and at once find ourselves staring in amazement. Whole families sit next to their gleaming vans with open trunks, with neatly laid tables—here and there even decorated with a candle—picnicking in their team colors. We enter the stadium and the match begins. Most players are black hunks who have fought their way out of poverty through football. Here, everything turns around money. Interesting game phases are replayed in slow motion on immense TV screens that have been set up to the left and right, but which above all else serve to show advertisements. Just when I am starting to understand the game, the signal for an intermission sounds. The breaks are timed to the second with an eye on the commercial messages.

It is more than ten years before the start of the Flemish Television Company VTM in Flanders, and I have barely digested the switchover from the National Institute for Radio Broadcasting to the Belgian Radio and Television Broadcasting Organization, the wildly popular Flemish sitcom *Schipper naast Mathilde*, and the arrival of color television. On Dutch television I cast half an eye at the advertisements, but Loekie the lion is a shy kitten compared with the powerful claws with which this country's consumer industry grips the land. Sport in America is TV, and TV is advertising.

Thursday, July 14, 1977

Life is busy at the CDC. I perfect the performance of techniques in a maximum security lab, describe all movements in detail, and create reports from the descriptions. Even writing is awkward in an MSL. I can never take my notes outside. I tape them to the window of the lab and then rewrite everything by hand on the other side of the glass.

Friday, July 15–Saturday, July 30, 1977

A family holiday in America. We visit Disney World, Cape Canaveral, the Rocky Mountains, listen to the blues in New Orleans, and savor the sea air that blows in from the Gulf of Mexico.

Sunday, July 31, 1977

Karl gives me a golf lesson. I require 121 strokes for 18 holes, just under seven strokes per hole. Usually you should be able to complete a hole in three or four strokes, but Karl laughs heartily because he discovers places on the course that he has never been before. Back in Belgium, golf will become a real hobby. With golf partner Rudi the Baron Verheyen, the honorary rector of the University of Antwerp—deceased at the beginning of 2014—I will further develop my skills in this fiendishly difficult ball sport.

Saturday, August 13, 1977

My family returns to Belgium. Aircraft and Dina: not a good match. This time they fly via Bermuda and there the flight is delayed for six hours. Fortunately, another group of Belgians is stranded with my family, which lightens the ordeal.

Tuesday, August 16, 1977

America mourns: The King is dead. Succumbed to a heart attack. Between the endless news reports on Elvis Presley's death, the networks screen ads. A heartbreaking interview with a fan is interrupted by a lady licking an ice cream, immediately followed by a report that speculates about an overdose of medications.

Wednesday, August 17, 1977

The air-conditioning of the MSL is kaput and the temperature climbs to 35°C (95°F). It resembles Yambuku.

Taking leave of Elvis grows into a national event. From all states people stream to Memphis, Tennessee, to bid their god a last farewell.

That evening Dotty speaks from her heart and says she is looking forward to Stan's return at the end of the month. Dina knows what it feels like to be married to an itinerant researcher, and then I am small beer in comparison with Stan. He is playing a key role in the eradication of smallpox. It is one of the first contagions that can probably be conquered by mass vaccination campaigns. Right now Stan is in Somalia, where smallpox is still claiming victims. An armed border conflict with Ethiopia threatens the work, but the exertions of people such as Stan will eventually pay off.

Tuesday, August 23, 1977

During the past few days we have grown large quantities of Ebola in cell cultures. I have taken a great many electron microscope photos of this. Never before have I seen such a clump of threadlike Ebola viruses. There are different ways to capture images of this pathogen. You can photograph it after it has come out of a cell and is completely free and alive. But you can also "fix" the cells so that the virus is neutralized. Then you can bisect the cell. In this way you obtain a cross-section of the virus or, depending on its position in the cell, a lengthwise section. It is fascinating to see the viruses at their maximum size. You obtain a pinpoint sharp image of the "tentacles" that the Ebola viruses use to fool a cell's doors into opening. I count myself lucky that I can be a witness of this.

End of August 1977

My sojourn in Atlanta is almost over. I have learned much, enough to construct a small-scale variant of a maximum security lab in Antwerp. I have also learned that things can go wrong. During the previous week we isolated viruses from cells that we had monitored continuously, in the hope of establishing a clear cytopathogenic effect. But after six days I see that the flask is completely filled with mold. Either we worked imprecisely, or we used fluids that were not sterile. To work off our frustration, Karl and I go golfing.

Late 1977

Back at the ITM, I report on my training in America and get the green light from Stefaan Pattyn to construct my MSL. I will shoulder the burden alone, because my colleague Peter Piot doesn't foresee a reward commensurate with the time-robbing work required and would rather apply himself to sexually transmitted diseases and their influence on mother and child.

Just like the Kinshasa example, the isolator that I build consists of an outer frame over which a cover of transparent plastic a couple of millimeters thick is stretched. In one area of the plastic film there are two holes that provide entry to sleeves that end in gloves. If I put my hands in these, my reach is limited to the length of the sleeves. For anything farther away, I need tongs. Moreover my arm muscles tire quickly. I had already discovered that in Kinshasa. Your manipulations then become harder and less precise. Now I equip the isolator with a support upon which I can rest my elbows, so that I can work for longer and be more stable.

The last thing we want is to release Ebola or other pathogens into the Antwerp air. I therefore ensure the isolator is under negative pressure. Even if I accidentally prick the plastic with a sharp object, nothing can go wrong. In case the motor of the absolute filter fails, the room in which the isolator stands is also held at a negative pressure. Electricity is required for this, and in a power failure the generator of the ITM will provide it.

Before you enter the room, you must pass through two airlocks. In the first you strip down to your underclothes. In the second airlock, you put on your work clothes and boots. The second airlock is also the darkroom, in which you can work with fluorescence. Your suit is re-used, but it cannot be taken outside to be washed. That must happen in the same airlock. At the end of a week you place your clothes in an autoclave and afterward pack them in a sterile bag, ready for the next use. Also important: do not drink too much on workdays, because you will want to urinate too much, and that means frequent undressing, dressing, undressing, and dressing.

Putting objects in the isolator or removing them—without running the risk that viruses enter the air of the room at any instant, let alone the streets of Antwerp—is a laborious business. On one side of the isolator there is a round opening from which a long plastic bag protrudes, this being bonded to the isolator in an airtight fashion. To remove something from the isolator, I put my hands in the sleeves and place the object in question inside the outward projecting bag from within the isolator. Now with my hands removed from the sleeves, I tie off the bag—with the object inside—with two fasteners. Next I cut the bag through between the fasteners, roll up the cut ends, and melt them closed with a bag sealer. The fasteners may now be removed. The object is now outside the isolator in a hermetically sealed plastic bag. Throughout the procedure the air in the room has remained separate from the air in the isolator.

To place something inside the isolator, I put it inside a new plastic bag. I pull its open end over the truncated end of the previous bag, which still adheres to the isolator. I fasten the new bag in an airtight fashion to the wall of the isolator. Next I put my hands in the isolator sleeves and pull in the shortened old bag, turning it inside out. Thereafter I remove my hands from the gloves and from outside I push the object from the new bag into the old bag. I again put my arms in the sleeves, open the old bag that now extends into the isolator, and remove the object, which is now inside my isolator. The transfer port has remained airtight throughout the process.

People who work in an MSL must marry patience and precision to nerves of steel. I keep that uppermost in my mind now that I must appoint a colleague. My choice falls on Greet Beelaert. In the five years that we will work together, not a single dangerous incident will occur in our MSL. No needle prick, no broken test tube, no smashed beaker.

Our MSL is inspected by staff of the WHO. They judge it to be as good as an MSL of level 4, the highest level, with the limitation that we must ensure that as few people as possible work in it and that I will never undertake experiments with any animals except baby mice. Having secured the approval of the World Health Organization, we are becoming known internationally in our little world. Gradually I am coming to be regarded as an expert in the business. Now and then an invitation to a reception or a congress appears in my pigeonhole.

Early 1978

Against the backdrop of the Cold War, an agreement for joint medical research by Belgium and the Soviet Union is signed in Antwerp. Now that dangerous viruses can be studied in our lab, I am required to share my knowledge with colleagues, such as those of the Institute for Poliomyelitis and Viral Encephalitis in Moscow, headed by Dr. Drosdov. Russians are warm people with whom it is good to cooperate. I teach my first Russian colleagues who come to the ITM how to work in an isolator and handle viruses such as Lassa, Marburg, and Ebola. Upon their return to Moscow they take samples of the live virus in thermoflasks. These pass customs because they are accompanied by a document that is signed by the director of the ITM in which it is stated that the "reagents" will be used exclusively for research.

Naturally, the thought crosses my mind that the cooperation of Belgians with the Soviets could be a way to gain access to information and even biological material that could also be used for biological warfare. But who am I to make known my worries about ulterior, improper motives for a scientific research program that has been supported by the government of my own country?

The ITM receives a letter from Peter Gregor Rytik, the director of the Belorussian Research Institute for Epidemiology and Microbiology in Minsk, White Russia. Rytik invites me to visit his MSL. He doesn't need to ask me twice. But I will have to acquire another passport for this. Americans would not like to see me arrive with stamps from the Soviet Union, and vice versa.

The White Russians have reserved a beautiful large suite for me in a relatively swish hotel in the heart of Minsk. Peter Rytik, a fascinating conversationalist who will later come to Antwerp on a return visit, shows me his lab. However deep the water is between the Soviets and Americans, I can't rid myself of the impression that the power mongers of the world, who could drink each other's blood so to say, have jointly settled how you must analyze that blood in a maximum security lab. Rytik listens attentively as I explain to him how we in Antwerp work in our small but efficient, WHO-approved lab.

Because Minsk was largely destroyed at the beginning of World War II and was thereafter rebuilt, the city has a modern appearance, with parks and wide streets. It would be pleasant to live here if there were more products on the shop shelves. For a bit of variety in food or consumer goods you need an extended network. I become acquainted with academics who work a kolchoz in their free time in order to harvest vegetables for themselves. They take home full bags for their own consumption or to exchange with, for example, a worker at a shoe factory.

One evening I have agreed with my colleagues to play a little cards in the lobby of my hotel. Our little company comprises about ten people. Some of them carry a heavy briefcase. They have probably done quick shopping. We toast our budding friendship with beer, Schnapps, and vodka. There is hearty laughter. Among the other people in the lobby the number of women is conspicuous. They are searching for a man in this city where even more than thirty years after the war there is still a surfeit of women. At 22:30 on the dot, all the lights go out and dim emergency lighting switches on. Around us I now see that most of the guests have left. The festive atmosphere being thus dampened, and the bar being shut, I stand up and signal that I am about to take my leave. But then someone says, "Who wants to drink another glass in Guido's room?" Now I realize why they booked a suite for me! Late into the night my "guests" smoke and gulp the vodka that they have brought with them in their briefcases, and eat bottles of pickled gherkins. Happily my suite has a separate room where I can sleep in a relatively smoke-free environment. My breakfast is the only bottle of Russian champagne remaining.

Hunting viruses immediately exposes a virologist to world history. During my stay in Minsk, I go to the Forest of Katyn, where an estimated 22,000 Polish officers and intellectuals were murdered by the Russian secret service during World War II.

During another working visit to the Soviet Union, I visit Zagorsk, a town not far from Moscow that is best known for its magnificent cathedral complex. When I was there for the first time, I did not know that there was an underground laboratory not far from the cathedrals. In 1999, the Kazakh Soviet doctor and former director of Biopreparat, Ken Alibek, would refer to the subterranean lab in his book *Biohazard*. When he still lived in Zagorsk, Ken Alibek—who was then called Colonel Kanatzhan Alibekov—busied himself with hemorrhagic fever viruses, among others. In his provocative book he claims that the Soviets had a program to grow viruses such as Ebola and Marburg on a large scale and then to freeze dry these into a powder form that could be loaded into rockets. From a technical standpoint that must be possible, but it demands the culturing of viruses on an enormous scale, which isn't so simple. When President Gorbachev came to power, the research program was canceled and Ken Alibek emigrated to the United States. Of course, he was thoroughly questioned there. Acquaintances of mine were among his interviewers.

In December 2014 I will once again renew my contact with my fellow filovirus expert, Dr. Jens Kuhn, and ask him what he thinks about Alibek's claims. Jens answers that the former Soviet Union did not possess filovirus weapons. According to him, the Soviets had indeed studied filoviruses, but they never managed to produce them in large quantities. Marburg grew somewhat better than Ebola, and in Kazakhstan a few experiments were conducted to see how Marburg was spread by explosions.

Regardless, it made sense that the Soviets would have wanted a program for biological warfare, if only because the Americans had one. The US even allowed information about its programs to leak to Moscow in order to intimidate the Soviet Union. In the meantime, the US had stopped their biological weapons program while the Soviet Union had pursued its own. The current status of Biopreparat is unknown. Alibek once again lives in Kazakhstan, richer by a heap of dollars, and has again adopted his original name. In 2012 Jens Kuhn will publish a book in which he lays many of Ken Alibek's claims to rest.

But of course I do not spend all my time in Russia. In Antwerp we had meanwhile grown a supply of live Ebola. Our stock consisted of plastic vials in which we had placed the virus using the isolator, and that we had subsequently melted shut. The vials lay in the deep freezer. But for diagnostic tests we require cells with dead Ebola and during killing the structure of the virus should not change. An appropriate way to inactivate the virus is to expose it to gamma rays. I learned in Atlanta that this can be done with a cobalt "bomb." They had such a radioactive source there, but for us that is unaffordable. I must therefore search for an alternative.

In the Walloon region of Fleurus we find a company that sterilizes operating research equipment with gamma rays. I ask the director of the ITM if I can drive to Fleurus with my live Ebola. He agrees. No one else is informed except Walter, the building manager and handyman of the ITM. Walter is also the private chauffeur of the director and a trustworthy chap. We drive regularly to Wallonia, he and I, following our own safety regulations. Thus we drive with two cars and, en route, we switch cars. If a stranger were to come to know what cargo we are carrying and devise malicious plans, we would like to make such villainy as difficult as possible. On one occasion Walter drives in front with the box containing fifty vials of Ebola. Another time it is I who climbs in behind the steering wheel of the "hot car." In the chase car there is bleach, a protective apron, and gloves. Should car 1 be involved in an accident, car 2 can immediately intervene.

It could have been foreseen: I receive an official invitation from the Bewakings- en Opsporingsbrigade (BOB*), Belgium's criminal investigation department at that time. Could I go to their office to answer a few questions? Someone who visits Russian laboratories as well as research institutions of the American army must surely have aroused suspicion in BOB. Now I have nothing to hide, so I go. The gentlemen place before me several names of Russian colleagues who have visited the ITM. Do I know these people? I confirm that I do. The detectives make references to the fact that we are in a Cold War and that I have contact with foreign researchers who are involved in biological warfare programs. Again I affirm that work is undertaken in both the Soviet Union and the United States with viruses such as Ebola and Marburg, and that I have been present during a number of these investigations. I sense a little frustration on the part of my interviewers. Are they hoping that I know of sinister plans? Do they suspect that I am helping the Russians with such plans? But I am standing on solid ground and cannot say anything more, because I don't know anything more.

For days after this remarkable "interrogation" Dina notes that a Volkswagen is continuously parked opposite our door. And when she lifts the telephone, she often hears a little click. Have we become paranoid, or are they really interested in me? And "they," is that the CIA, KGB, or merely our friends at the BOB?

Late 1978

I receive a request from Japan to give my opinion about a new maximum security lab in Tokyo. The Japanese wish to show it to the press shortly. So it's van der Groen onto the aircraft.

My colleagues clearly haven't seen a yen. This seems to be the Rolls Royce of MSLs. The isolator of gleaming stainless steel is complemented by thick plexiglass and equipped with tough sleeves and gloves. The director of the lab stands beside me and glows, as do the engineers to the left and right of us. The lab elicits the comment from me that it is the most impressive MSL that I have ever seen, but that I can only judge it if I have worked in it. The isolator intrigues me, because there is an ultracentrifuge standing on its floor! It is amazing that they could build such a piece of equipment into the isolator. The rotor of an ultracentrifuge attains crazy accelerations, requiring it to be large and heavy. I put my hands into the sleeves and immediately feel they are not good. At six feet I am a good deal taller than the average Japanese, but quite apart from that, the sleeves seem to be wrongly made. When you design an isolator you must ensure that a researcher using the gloves can reach as many objects as possible. I try to open the lid of the ultracentrifuge on the bottom of the isolator. I manage this, but with the best will in the world I cannot raise its rotor. Frustrating, of course, because it is in the rotor that you must place your centrifuge vials. Cautiously I suggest that something is not quite right with the dimensions... I could just as well have dropped a bomb. The lab falls horribly quiet and the engineers turn all colors of the rainbow.

The Tokyo MSL is not opened, but that cannot be ascribed to my observation. Surrounding residents will not have it.

May–June 1979

I become involved in the WHO's plan to eliminate smallpox by means of large-scale vaccination campaigns and expeditions to all corners of the Earth where it is suspected that the virus still lurks. It is an ambitious plan, complicated by the fact that in Africa the virus can be confused with monkeypox. Primate poxes can also infect people, and at the beginning the disease resembles normal smallpox but the course of monkeypox is milder. A definitive diagnosis can only be given by blood analysis in a lab.

Since the beginning of the year, the WHO knows where to find me and the organization asks for my advice with regard to hemorrhagic fever and laboratory safety. The WHO also knows that science—and here the ITM is not in last position—is still searching for the animal host of Ebola. From this it transpires that I may go with an expedition to the area around Tandala in Northwest Zaïre. For the ITM it is doubly justifiable: we will make a contribution to the eradication of smallpox and we can investigate the as yet unknown animal reservoir of Ebola. I will be away for two months.

Before we travel to Tandala, we spend a few weeks in the mission of Yalosemba. Once more I am a member of an international team. My CDC blood brothers Joel Breman, Dave Heymann, and Karl Johnson are again in the party, but I become acquainted with some new people: the passionate American brothers Brian and Lynn Robbins, both zoologists in the Smithsonian Museum in Washington—one of my favorite places in the US—and Mark Szczeniowski, an American who has already worked for many years for the WHO and who appears to be an organizational superman. In addition our company has two Zairese researchers: the zoologist Oyndole and the doctor Kalissa Ruti. Just as three years before, Karl Johnson is the leader.

There have been cases of Ebola in Tandala, but this time we are seeking infected animals that carry the virus and do not die. Bats are the usual suspects, because the related Marburg virus has already been found in this type of animal. Above all else we want to obtain specimens from animals. We will not do testing this time. That will happen later, in the laboratories of the CDC and our own facility in Antwerp.

On this occasion, too, we enjoy the hospitality of the sisters: Sisters Ghiselaine, Merthildis, and Amandina, who are very friendly. Further, the Scheutist Father Willy Daem bids us welcome. Willy is a sympathetic man who speaks fluent Lingala and who wants to bring us into contact with the village chiefs. We had noted in Yambuku that this is very important. In this way the chiefs know your plans and as a researcher you learn a great deal about local practices. To find the Ebola reservoir we need many different animals. On our own we can try to catch bats, but for the other kinds of animals we must rely on the help of the locals.

I will keep a verbal diary with the aid of a cassette recorder that I loan from Mbabanza, a Zairean nurse who was trained by Sister Romana. Until September 18, 1976, Romana was the head of the mission at Yalosemba. Then she departed for Yambuku in order to replace Sister Beata who had fallen ill. Romana was herself infected and died two weeks later. One of the few worldly goods that she left was a tape recorder. Her family donated it to Mbabanza and now I am using it. A strange but beautiful circle.

Father Willy and I visit the villages in the neighborhood, armed with a heap of animal photos. We inform people of our intentions and note the native names for bats, porcupines, and rats. We ask what the people like to eat and probe carefully to find out which animal species are associated with superstitions. Requests to catch an animal that is loaded with superstition would only stir unrest. For example, certain animals are believed to have a negative effect on the unborn child of a pregnant woman. Eventually we have a list of target species and for every animal we set an agreed payment. In discussion with Willy we determine how much rice, sardines, meal, or zaïres a trapped animal is worth.

In one part of the church that Father Willy can spare, we improvise a dissection room. Each day eight hunters venture out, accompanied by two porters and a technician. The porters carry the bag as quickly as possible to the Land Rover, which transports the animals to our lab. The technicians are useful youngsters that Mark Szczeniowski has trained to conduct interviews and to give vaccinations for the smallpox investigation, and to track down cases of monkeypox. Now, during the hunt, they will control the ammunition supply and draw blood samples from the hearts of the animals caught. They must transfer the blood to little vials, for each of which they must provide a fully completed label. Not always so simple, in the middle of the jungle. To be honest, we have little faith in the whole setup, but already on the first day Karl Johnson and I see about ten animals lying before us: monkeys, bats, and a shrew. We begin dissecting at once.

Sister Ghiselaine departs for the Yambuku mission. We ask her to take a request to Yambuku: would it be possible to once again draw blood from the people who took part in the plasmapheresis program of 1976?

Hence we will be able to check how their antibodies against Ebola have changed after three years.

One evening I accompany the hunters. The bats skim the water surface, searching for a meal. Over the whole width of the pond we have stretched a net one and a half meters high, as if we were going to play volley ball. Our ruse works. The humid evening air attracts swarms of insects to the pool, where they end up in the small gaping mouths of the bats. Echo-location is a wonderful system, but their sonar cannot spot our fine mist net. I am a little heart sore when I see how their flight is abruptly and silently cut short. And then little Guido appears before me. He wants to haul his young friends on board his imaginary aircraft. My childhood dream to be a pilot never materialized, although I had weighed it. Happily, my life has turned out such that I fly often. And now after yet another flight I stand here in the Zairean jungle, catching bats.

A stabbing pain shakes me out of my daydream. Wearing only shorts, I paddle barefoot in the shallow water. I feel how worms bore under my nails and into my toes. I stifle my discomfort because it is time to disentangle my prey from the net.

With time my hunter's instinct blossoms fully and I become much more skillful at catching bats. I catch them in all their various forms: from powerful fruit bats, with heads like apples, to tiny versions, two of which fit on my palm. These potential Ebola carriers—as we must regard them—all have razor-sharp teeth. We must therefore handle the animals warily. I wear thick gloves but that is it with regard to protection. There is no hint of the spacesuits or hermetically sealed hoods that will appear daily on TV in 2014. You must try it once, disentangling a struggling bat from a net and putting it in a cotton sack with oversized gloves! We carry the sacks to our lab in the church, where we hang them neatly on metal hooks that we have screwed into the wall.

One morning I wake up at 6:00 after a badly disturbed night. I kept on hearing a sound, the source of which I could not identify. Well now, against the window sits a whopper of a locust, at least 10 centimeters long. Prey for the bio-brothers Robbins. Sister Ghiselaine is back from Yambuku and brings good news. Sister Geno is departing for Belgium on holiday and will let Dina know how things are going with us here.

After breakfast I set to work in the lab with Brian Robbins and a local assistant. It is my task to take tissue and blood samples from every bat. By trial and error I succeed in determining the correct dose to anaesthetize the animals. In front of me is the plank to which I pin the bats. I disinfect the ribcage with alcohol and then cut the chest open with surgical precision. To obtain a proper blood sample, I push a needle directly into the beating heart. I am extremely aware of the risk and thus work calmly. In the background "A day in the life" from Sergeant Pepper's is playing. This cassette has lain too long in the sun, I think. There is a cyclical distortion of the music, although this fits the circumstances well.

For the current generation of researchers it seems unthinkable, but our safety precautions are minimal. In the humid heat we wear no more than underpants, a mouth mask, and a protective paper apron, although this has long sleeves with wrist bands that close neatly around our latex gloves. It is always suffocatingly hot here and in no time at all my apron sticks to my body like a mop on a floor. The Bunsen burner that I use to sterilize the scalpel blade with which I cut away parts of the liver, lungs, and kidneys, adds some extra heat. Perhaps the assistant has allowed himself to be transported a little too far away by the Beatles, because suddenly a sleeve of his apron catches fire, and it immediately flares up like a blowtorch. He reacts coolly, beats out the flames and suffers no significant wounds. So it is that colorful tales are born that you can recount years later with a beer in your hand and a wry laugh.

Because the blood and tissue samples could contain Ebola viruses, the form of which we wish to preserve for as long as possible, the specimens must immediately be put in a container with liquid nitrogen. We must monitor everything, and we therefore conceive a system using nylon stockings. In this remote corner of Africa one seldom seems women wearing pantyhose, but for a virologist they come in handy. I place each sample in a little vial with a preprinted label. On a form I stick a duplicate label. Alongside it I note the name of the animal species and the dissection date. The adhesive on the back of the label freezes and fails at -196°C, so I wind cold-resistant tape around the vial. I place ten of the little vials into a single plastic bag, according to their code number, and put the bag into a nylon stocking. Just above the bag I tie a knot in the stocking, so that everything remains in order. Then I deposit the nylon stocking in the container with liquid nitrogen. Nylon is thin, elastic, and can resist extreme cold. I have tied a cord with a numbered tag to the stocking, and the tag dangles outside the container. Operation Nylon Stocking appears troublesome, but it is the best procedure I can think of to guarantee a correct follow-up of the samples later. The container will soon be loaded into the trunk of a Land Rover to be carried over bumpy roads and thereafter transported in the belly of an aircraft. But with the sample bags trapped in the nylon stockings, little can go wrong.

Our animal collection steadily expands. Each day about twenty specimens are added. The local population watches our activities with increasing amazement. The people are wondering why we go to all this trouble, instead of simply throwing all these delicacies in the pot and gobbling them. They subsist on a diet of fruit and rice. At the market a piece of meat costs ten zaïres, not a small amount for people who earn one zaïre per day.

I suspect that the locals and certainly the hunters could do with some pep talk and decide to address them from the pulpit in the church. As a young teenager I found it was no fun attending mass, but on the preacher's stool I finally realize what it must have been like for the pastor. In the little bit of Lingala that I have picked up, I try to motivate my beloved believers to cooperate with us and catch as many animals as possible. I explain with hands and feet that we are seeking the host of a dangerous virus that has also made some of their fellow villagers sick, and which even caused a true epidemic three years ago. Also I propose that from now on the meat of the captured animals is cooked in a big pot and shared among the villagers.

The next day the hunters have already departed at 3:30 in the morning and by the time I am well and truly awake, a young antelope lies on my dissection table. The Robbins brothers are especially interested in skeletons. After I have dissected the animal, we remove the hide and innards. The meat goes in the pot and we let the rest of the cadaver dry in the sun. In the tropics that means party time for scavengers. The smell is unbearable, but skeletons are in the meantime eaten clean. Thereafter we can measure them and make them ready for transport to the United States, where rodent specialists will identify the species. The animal's hide is also preserved, because its shape and markings can aid identification. In this way some new animal species will even be discovered.

More so than in Yambuku, I realize that in Africa uninterrupted work days are unsustainable. This time I break off for a rest now and then. Then I do nothing more than look around and try to adapt to the ever-present red dust and the all-pervading, clinging smell of wood fires. I observe how women walk with water cans balanced on their heads while carrying a jerry can in each hand. With my dinner in the evenings I drink a beer. Primus beer is one of the safest drinks here. It is brewed in a relatively sterile environment and it suffices to hold a bottle against the light to know if it is drinkable. Murky is bad, clear is safe. Now and then a whiskey or two also goes down the hatch. This is okay, due to the high alcohol percentage. And then the emotion can kick in. Like now. In the distance I hear someone playing a trumpet, an unearthly sound in the jungle.

After a good two weeks at the Yalosemba mission I have harvested blood and bits of livers, spleens, and kidneys from more than one hundred animals. That day the hunters bring in 33 animals, especially bats and monkeys but also a rat and a porcupine. At 12:00 sharp Father Willy rings the bell for lunch. Discussions during the meal are always pleasant. Willy talks ceaselessly about the languages and dialects of Zaïre, which number at least 45, according to him.

It comes with the time of the year—it is well into the rainy season—but a couple of hours later, when the sun has passed its zenith, I am plagued by little biting flies that cannot be beaten away. They have cashed in on my arms and neck, and I am covered in lumps, which itch terribly.

Two days later I hang over the toilet continuously, vomiting and with persistent diarrhea. A pity, because Karl and I had made plans to drive to Yambuku and find Sisters Geno and Marcella. That must be called off. And nothing will come of this plan later. I have work enough here and perhaps I will not be able to make enough progress. Yambuku 1976 is buried in the past. It was an unusually intense episode and as a virologist I will often reflect on it and speak about it in lectures, but for myself the chapter is now closed. To combat loss of water and salt I drink Oxo. I don't worry too much, because up to now I have always worked safely, and apart from the sleeve that burst into flames, nothing has gone wrong in the lab.

When I have picked up again, seventy animals are lying there, waiting. In the interim Karl Johnson has left Yalosemba. His wife, Patricia, will arrive in several weeks to replace him. From now on I am the leader. It is the first time in my life that I lead a team and this role, I quickly note, I undertake with pleasure.

At the beginning of July we travel on to Tandala, a journey of twelve hours. We repeat the now proven procedure: mobilize the local population, inform them, and pay them to help us hunt. Mark Szczeniowski leaves for Gemena to fetch supplies—toilet paper among other things. The logistics and organization are now in the hands of Helen Engelman, who, coming from America, forms a tandem with Mark. If at any given moment I have too many onlookers in my lab, Helen solves that immediately. She arranges for a stout rope to be sent from Atlanta to Tandala, so that I can fence off my lab and keep the locals at a distance. Among today's animals there appear to be three monkeys infected with the monkeypox and whitepox viruses. In some rats we find severely inflamed livers, possibly due to Ebola.

That evening the generator fails. I hear the crickets louder than usual. It was hardly a romantic coincidence that I was busy putting my nylon stockings into the containers of liquid nitrogen. I will have to work overtime and literally fumble in the dark. Fortunately Mark is back and, besides toilet paper, has brought wine. With its aid the frustration is quickly washed away.

On July 12 we dissected the thousandth animal of our expedition. That day Patricia Webb arrives to replace Karl. She travels in the company of Mr. Mbaki of Zairean nature conservation. I await them in Gemena and together we complete the three hour trip to Tandala. I have last seen Patricia during my training in Atlanta, and we have a hearty reunion. In the evening we see an infinite heaven full of sparkling stars. Every star is an Ebola-positive case, we think—but perhaps that can also be ascribed to the whiskey. Alcohol and the jungle go well together—sometimes too well. In the evenings there is nothing better to allow you to relax than endless after-dinner prattle. Of course a little drink is in order, one which not only tastes better than water but which is also safer microbiologically. Closing hours or alcohol laws are irrelevant here, so what holds you back? Before you know it, drinking is habitual. Personally, I haven't developed any problems, but I have met enough other expats and tropical travelers who have drunk their livers to oblivion.

For the local population, Primus beer is much too expensive, let alone gin and tonic or whiskey, but they have their own brews. On one occasion one of our chauffeur-mechanics, an excellent chap mind you, has drunk too much nsamba, a kind of palm wine. He misses a bridge and runs the Land Rover into the river. A colleague pulls the Land Rover out with one of our four wheel drive vehicles, onto the riverbank, where it can dry out. On the way back to Tandala our cowboys race so quickly over the bumpy roads that the second vehicle also packs up. A third Land Rover was put aside earlier with a broken alternator. And thus within the blink of an eye we find ourselves without a vehicle. Patricia must visit the surrounding villages, but she doesn't get riled and simply uses a bicycle.

As the expedition has almost finished by the end of July, we make our equipment ready for transport. We are certain that we have been able to guarantee the integrity of the cold chain. Everything depends on this. The specimens that we have collected must be placed in liquid nitrogen at -196°C as quickly as possible. To have enough of the stuff at all times and places we had to call on diverse suppliers but that went well.

A military cargo aircraft will take us to Kinshasa. The young pilot comes to me and asks if I am the team leader. I affirm. He nods his head in the direction of the Land Rovers, besides which there are heaps of crates and boxes. "Consider the fact that I also have cargo, so try to pack everything as densely as possible," he grins. I am instructed to arrange everything in the belly of the aircraft and must also estimate how many tons the freight weighs. As soon as the Land Rovers have driven aboard, we strap them securely and try to stash the rest of the cargo around the vehicles. Then we sit down at the rear of the plane, on crates that serve as seats. Not much space remains. I have completely forgotten the pilot's freight when he whistles through his fingers and I see about ten heavily laden men approaching the aircraft. They are carrying cans and baskets with live chickens. A little goat also skips among the company. Man and beast squeeze themselves between the freight and the wall of the aircraft. The crew takes it all in their stride. Their jovial mood betrays the fact that they have downed a good few Primus beers.

We leave. Through one of the few windows next to me I see a dust cloud swirl upward, and in the background a menacing sky and lightning bolts. If only that settles. It takes forever for the kite to leave the ground. Just ten meters of runway remain when we finally lift off. During the following flight of ninety minutes we experience terrifying turbulence. The airframe bucks so violently that some of us literally fly into the air, falling back a meter away. People scramble helter-skelter to avoid flying objects. I look anxiously at the straps that must keep the Land Rovers and containers in place. Everyone is silent; one cannot fight against the roar of the engines. Eventually we land in Kinshasa, equipment intact, where the thunderstorm has made way for a rainbow of relief.

How in God's name can eight adults and all the luggage and crates fit in the Volkswagen microbus that is waiting for us? But we manage. We will fetch the fully loaded Land Rovers later. On the way to the Memling Hotel we are stopped by soldiers. They want to check our luggage and ask how much money we have. We have no interest in this. We reason our way out of the situation with a small fee and the bunkum that we are on our way to fight cholera on the orders of the WHO.

Later, from a bay window of the hotel, we see how policemen set a trap. First they let motorists drive down a street to a point where a police car blocks the way. Two powerfully built men with sunglasses and a gun show the drivers a "no entry" sign behind them, which a colleague in the conspiracy has in the meanwhile placed at the start of the street. The fine is levied and the traffic sign once again removed, to let in the next lot of offenders...

Finally we are in the Hotel Memling and a little later our two Land Rovers also stand at the door. We begin to remove a few chests with valuable research equipment from the trunks and pile them in the lobby. Suddenly Lynn Robbins comes running to me. The crate with skulls and skeletons has disappeared. But we do not panic and decide first to grab a bite to eat and to let our travel-battered brains recover. During the dessert we hear that some youngsters are loitering around our cars. We give a little speech, explaining that the contents of the stolen crate do not contain any food but rather an evil virus that has cost the lives of many people. We promise to pay the genuine finder 75 zaïres, and then eat the rest of our dessert. Not even half an hour later the crate is back, unopened.

From Kinshasa we fly to the headquarters of the World Health Organization in Geneva to report to Paul Brès. We want to secure a follow-up mission. We have captured many animals, the cold chain remains intact, and the composition of the international team was well-balanced. With 27 specialist fieldworkers—biologists, virologists, doctors, technicians, and drivers—we sought three viruses: Ebola, monkeypox, and whitepox. We have worked outstandingly well together, with humor cementing all the nationalities and specialties. At the leave-taking function I thank everyone at length. I have enjoyed the expedition and now hope above all else that in the maximum security lab of the ITM we will find clues that will unmask Ebola's host.

August 1979

We study the many hundreds of animal specimens that we have gathered during the last few months. The stakes are high: if we can find the reservoir of Ebola, we can advise populations expressly not to eat this animal species and to treat it with the utmost respect. We search for living Ebola virus in the specimens. To do this we grind the pieces of organs and tissue finely and introduce them to vero cell cultures. I follow the culturing process to see if cell damage appears. It is stressful and time consuming work that takes weeks. The disappointment is hard to swallow when, after two months of intense work in the African jungle and more exhausting weeks in the ITM, we find nothing. Not one animal was carrying Ebola. It is a defeat for science and for myself.

September 12–October 8, 1979

In the South African city of Johannesburg a new maximum security lab is opened. The global Ebola community is invited to attend the formalities. This time Dina accompanies me. The land is still in the midst of apartheid. It already starts in the airport. A separate exit for whites. In the bus we sit at the front, among the other whites. But it is wonderful to see colleagues such as Margaretha Isaacson again.

For the first time, Dina and I go on safari, in the Kruger Park. Karl Johnson has a camera with him which has a telephoto lens so long he could even snap impalas on the moon. We have a fantastic time, which makes such a deep impression on Dina and I that we will often go on safari during the rest of our lives. We are forever grateful to the organizer Walter Prozesky.

Sunday, December 9, 1979

The world is declared smallpox-free by a group of eminent scientists. The last case was reported in Somalia. Several months later, on May 8, 1980, the conclusion of the scientists is confirmed by the World Health Assembly. It is the first time that people have eradicated a dangerous virus globally. It has cost years of work and perseverance, and as far as I am concerned as a human achievement it ranks right up there with the moon landings. Samples of the living smallpox virus are preserved in two places: in the CDC in Atlanta, and at VECTOR, the state center for virological and biotechnological research in the Russian town of Koltsovo.

We have almost managed to eliminate polio too—at one instant the virus was still present in just Pakistan, Afghanistan, and Nigeria—but through armed conflicts and political instability or lack of will, this has not been achieved. Disinformation also plays a role. In Nigeria and Pakistan vaccination workers were killed because people suspected that they wanted to sterilize Muslims or inject them with HIV. Bill and Melinda Gates support the campaign against polio and hope to see this disease swept from the world by 2018.

February 1980

A thick mist blankets Zaventem and again I take my leave of Dina. She has an acute cold and makes me swear that I won't stay away longer than three weeks. This isn't my plan either, although in Africa you can only know when you are leaving upon actual departure. The WHO is still searching for the reservoir of Ebola and this time I am traveling to Mbatika, a pygmy village in Cameroon. My American colleague Dave Heymann is already in that area. A year earlier Dave investigated blood samples in Mbatika and determined that 17 percent contained antibodies against Ebola. The population must have come into contact with the virus, but either they do not fall sick, or else there were just a few patients and their symptoms were confused with those of another tropical illness. The suspicion is growing that there are also milder forms of Ebola. Maybe these are the rule, and the death of 80 percent of infected people in and around Yambuku in 1976 was an exception? Whatever the case, the question remains: in which animal species does the virus lurk between epidemics?

On the aircraft I read parts of a book by Jean-Pierre Hallet, a Belgian anthropologist who studied the society and activities of pygmies for thirty years. He even once lived in a pygmy tribe for eighteen months. According to him pygmies are the oldest representatives of the human race and of all people they are the most closely related to Australopithecus. Pygmies, declares Hallet, are courageous and musical, are of a good disposition, do not know war, and believe in one god—creator of the world and of man. I am intrigued.

On February 14 at 22:30 we land in Douala, Cameroon's largest city. The sultry air feels like a warm hand on my winter skin. Dave Heymann is standing there waiting for me, a broad grin on his face. After a bear hug he lets me know that he hasn't quite finished the preparations for the trip. Yet more equipment must arrive from America and we only have one car for nine people and a heap of luggage. Meanwhile, I am thinking about the two containers with 60 liters of liquid nitrogen that I must steer through customs. Fortunately, Dave has a letter from the Cameroon Minister of Public Health that states that the containers are of inestimable value for an important expedition. That document and a 1,000 CFA banknote work wonders.

I am settling into my hotel room when there is a knock at the door. I open it and find myself staring into the eyes of an African beauty. "Bon soir, je me sens seule. Et toi?" I search for civilized French, but my tired stammering apparently speaks volumes because the lady doesn't even wait for the end of my sentence and knocks on another door farther on.

The following day, Dave has organized a Toyota Landcruiser, plus driver. They safely take us 295 kilometers farther, to the capital city of Yaoundé. There we stop at the building of OCEAC, a French organization for development aid that cooperates with the CDC. In expectation of the equipment that must still arrive from the US, we load several metal coffers full of goods from the supplies of OCEAC's small laboratory. Dave introduces me to Ethleen Smith, his friend and medical assistant, and to Jean-François Bergman. Jean-François is a Parisian doctor who is the spitting image of Johnny Halliday. He and Ethleen will accompany us to Mbatika, our final destination. Simon van Nieuwenhove, who was a part of the Yambuku team with me, also joins us.

The other car that Dave has managed to secure is a dilapidated Peugeot that will be driven by Douglas Palmer of Aid International Development. Douglas is a good friend of Dave's, well-muscled and with a tanned, tropical countenance. A trusty comrade in the wilderness, it would appear.

At dawn the next day, Dave and the others have already departed in the Peugeot. Simon and I are still busy loading the Landcruiser with equipment, in expectation of the two expedition members that must arrive shortly. Suddenly I hear a familiar voice behind me: "What's a Belgian doing here?!" It is Patricia Webb, who has flown in from Sierra Leone. The reunion with the 55-year-old American is joyful. In Atlanta and Tandala I learned to value her sharp mind and sense of humor greatly. Patricia has brought along biologist John Krebs. The Americans have endured a night journey of ten hours with six (!) landings en route and are now gulping strong coffee.

It is blood hot when a few hours later we begin a journey of 390 kilometers, with Simon driving. The traffic is dense and the road dusty. For every oncoming vehicle we have a ritual: close the windows tightly, switch on the windscreen wipers, let the oncomer pass, open the windows, switch off the wipers. But the red dust penetrates everywhere, between my teeth, in my throat. Gradually we come to just leaving the windows open.

As it gets dark, we drive through a vast forest. In the headlight beams the trees—which I estimate must be a good fifty meters tall—look even more impressive. We stay overnight at the mission of Bertoua, where Dutch missionaries who speak French show us to our rooms. There is a mosquito net, a refreshing shower, and no rumble of generators. Bertoua is a town with an electricity grid. We savor the luxury and anaesthetize our aching throats with whiskey. When I am lying in my bed and turn the light off, I see a luminous picture of Christ. How well I will sleep tonight!

After a breakfast with fresh baguettes we resume our trek. At the market of Batouri we have a short stop. The two women disappear into the crowd and wander back a little later with bottles of wine and a bundle of rolled tobacco. According to Simon, the pygmies are crazy about it. After midday we reach the mission of Yokadouma, where two Belgian missionaries welcome us. The Flem Stef Wuyts and his Walloon colleague Jules Perpête live there as brothers with Hyppoliet, a Cameroonian priest in training. We chat into the wee hours.

In the morning we make the acquaintance of a French nun who has arrived from a nearby mission. Her name is Noëlle and she asks if she may accompany us. Sister Noëlle is a doctor. In Paris she specialized in radiology and has since been in Cameroon for two years, works in a dispensary, and already speaks Baka to a useful degree. I ask what motivates a young woman of 35 to undertake mission work so far from home. "I feel that God has chosen this place for me," she replies. This passes right over my head, but Noëlle seems perfectly happy.

We are sardines in the Toyota, but Noëlle can still squeeze herself in between Simon and me. The journey goes well until Simon suddenly loses control of the steering and the car skids off the road. We come to a standstill in a ditch. The Landcruiser is tilted dangerously downward on my side. I am buried under Noëlle. With much trouble I extract myself from beneath her and worm my way out of the all-terrain vehicle. It has a flat tire. Together we manage to level the vehicle and begin to change the tire.

The road we've veered off of has just been laid. Over a width of at least forty meters bulldozers have felled trees to make way for the road. When you rip such a mouthful out of the wilderness, of course you disturb the entire ecosystem. It is also an effective manner to facilitate the transmission of zoonoses* to humans. Animals that would otherwise never come into contact with us now have a spacious track and spread bacteria and viruses. New roads such as these also change the traditional lifestyle of the pygmies, explains Noëlle. Of the estimated 40,000 pygmies in this area, more than half have had contact with mission personnel.

"They are an ancient people," I say. "Their way of life has been tested and moreover it has been proven. Perhaps we would do better to leave them alone." But according to Noëlle, it is much too late for that already, and the customs of the pygmies are irrevocably "contaminated." Their traditional monogamous relationships are yielding to the polygamy practised by most other tribes. Their primitive little huts of palm leaves, built in a flash, are being replaced by sturdier structures that tie this nomadic people to the ground. The hunter-gatherers of former times now cultivate cacao and mangoes and bury their dead bodies, which in earlier times they left to the vultures. We find burials and all that goes along with this to be a sign of civilization and respect, but if life is your love, one would do better to act as the pygmies did in previous times. The less contact one has with a dead body, the smaller the risk that you will suffer the same fate.

In Yambuku there were two big fire accelerants: over-zealous injections with contaminated needles and funeral rites. Bodies were touched extensively by many relatives, washed, and kissed. We have seen the result.

On the way to Mbatika we have just one more stop, at a French mission. We meet the Mother Superior Marie-Laurence, who has already lived thirty years in the tropics. Her young colleague Marie-Claude has been here for just four months. Marie-Claude has a pleasing appearance and somewhat resembles Mia Farrow. But I ask no questions about her calling and limit myself to silent regret. There is also a pygmy couple. I shake the hands of the spouses and gaze into their eyes. These people are about one and a half meters tall, but their age is more difficult to estimate. I tell Mother-Superior about the book of pygmy expert Pierre Hallet that I read on the aircraft. "Ah, ce charlatan..." sighs Marie-Laurence. She knew him personally, she says, and recalls that above all else he wished to be famous. To achieve this he used the pygmies. Something to mull over during the last stage of the journey.

We drive into Mbatika and see Dave's old Peugeot already standing there. They have only just arrived but have covered the thousand kilometers somewhat faster than ourselves, despite five tire punctures and several stops for a recalcitrant petrol pump. Dave introduces us to Sassi, the chief of the pygmy tribe. He proudly shows us the spacious hut with four small bedrooms that the villagers have built for the expedition. There is even a lab space: four stout poles under a sloping roof of palm fronds. Then Sassi invites us to accompany him a short distance into the jungle. There they have dug a pit that is a good four pygmies deep and roofed it with a couple of thin tree trunks. Our toilet.

Must we hunt for insects, animals that frequent human dwellings... or yet again bats? With my equipment list in one hand and the experience of previous expeditions, we construct a primitive lab on top of the tables that we have brought. It appears that our supplies are short of one table. In nearly 20 minutes, Sassi and some village companions make an extra table from light and stiff bamboo. Biologist John Krebs hangs his spring balance from the ceiling. Done.

Ethleen, Patricia, and I will dissect the animals. Dave and Douglas trek to the villages for our advertising campaign with Vincent, a pygmy who can read and write and who speaks French. People who wish to hunt for us will be rewarded with salt, soap, money, or gas. The pygmies do not need firearms, because they are so accurate with spears and crossbows.

Almost a week ago I sat in my small but hypermodern MSL in Antwerp, and now I am standing in the jungle in Cameroon cutting open animals, among fascinated pygmies. Their eyes are riveted to our activities. White men and women in yellow aprons with long sleeves, rubber gloves, and a blue mask over the mouth measure animals, weigh them, and peek under the tail to determine the gender. Like a madman, one of them scribbles everything on a few sheets of paper under different headings. The other dips a laughably small knife and other gleaming implements in a liquid from a bottle which has a picture of a skull on it and subsequently holds them in a blue flame. Then he presses one of these little knives onto the belly of an animal and cuts it open carefully. When I produce my nylon stockings, my fans' eyes leap out on stalks. Because I only need the legs, I cut them off. I give the transparent briefs to Sassi. "Here, for your women," I wink. I open the container with liquid nitrogen. A plume of white steam curls upward. The pygmies are stunned.

Animals are food—that we know—and so I give Sassi a newly dissected pangolin as a gift. He thanks me and at once gives the scaly animal to one of his wives. She grips the armored beast by the tail and places it in the little fire in front of the hut. I hear women making the typical Baka throat sounds. A girl throws reddish, hairy caterpillars into a pot on the fire, rolls a few green leaves together, and wraps these around the handles of the little pot to allow her to toss the hissing creatures into the air. She grasps one, nibbles it, and tosses the pot once again.

Patricia and Ethleen have gone fishing. For pygmies this is women's work. In the case of our researchers there is still room for improvement. Hours later Ethleen returns with one unsightly little fish and Patricia with a foot infection.

I carry out the last dissections of the day by the light of kerosene lamps and make ready for the night. I think about Dina; it's her birthday today. I will be able to celebrate it at a distance because coincidentally there is dancing here. In a simple yet comfortable reclining chair that the pygmies have made, with a cup of instant coffee in one hand, I observe how the drummers strike up a hot blooded rhythm around the campfire. One man in a rustling skirt performs a dance to ward off illness. Women and children with little bells around their ankles join him. The drummers turn the volume up. After a few minutes all the dancers are dripping sweat. I cannot remain sitting still and let myself go completely. The children laugh a lot and imitate my clumsy movements. Out of tiredness, not embarrassment, I accept the invitation of a drummer who offers me a tom-tom—willingly, because I am nuts about rhythm. My uncontrollable urge to tap on anything and everything drove my mother crazy too. I rapped with my little spoon on every object that came in sight, when in the bathroom too! Even now I still grip a bongo between my thighs from time to time. But here and now, under the stark magnificence of an infinite starry heaven, beside a crackling fire and with hypnotized dancers around me, I truly feel for the first time the effect that music and rhythm can have on a mortal. Sassi looks at me, intrigued by how I lose myself in the cadence. "Yes, yes," I see him thinking, "Master Guido is in the mood."

Since the pygmies became acquainted with the white man, dress habits in the village have changed radically. Previously the people wore a loincloth, now they clothe themselves in T-shirts, shorts, frocks, and—very conspicuously—headgear. From borsalinos, through a checked cap with pompoms, to a wide-brimmed Canadian mountie's hat. Chief Sassi earns the crown of the jungle catwalk with a bowler. In addition he is wearing a yellow shirt, Bermuda shorts, knee-high socks, and army boots.

It is not easy to find animals. It is the end of the dry season and now the game shelters deeper in the jungle. The hunters must cover great distances. Our bag up to now is about twenty rats, a few monkeys and antelopes, one pangolin, one hyrax (or rock rabbit), and two tortoises. After hours of talking we find a hundred of the 120 village residents from whom Dave had drawn blood two years before. Names are a problem. People often give a new name to a stranger. Identification is nevertheless important, because if we find pygmies who now have antibodies against Ebola and they didn't two years ago, this indicates that the virus is still circulating.

While driving to a village in the neighborhood to fetch animals, we see two young white women walking beside the road. Simon stops and we engage in small talk. The clichés are affirmed: the chubby blonde girl appears to be an American and the slim one with her thick lenses, blue socks, and brown sandals comes from England. They too belong to an international team, which is translating the Bible into all possible languages. They have been living for more than half a year in a little more distant pygmy village in order to learn Baka. That is some job, I think, but maybe Cameroon has priorities other than a Baka Bible...

A little later in the village I hear a horror story about a Belgian missionary who is currently trekking through the region to vaccinate pygmies against measles with just one syringe and one needle! Yambuku squared. We do not learn from our mistakes. The good man must have made much mischief, because his bad reputation precedes him. In some villages the people flee into the jungle when they hear that the father is approaching.

It is Sunday, 7:00 in the morning, and still deliciously cool. I start with the dissection of five pangolins and one big rat. After midday Dave brings together some of the village residents. He wants to train at least two pygmies how to collect blood droplets on filter paper using the finger prick technique. He explains that the two must visit villages regularly to take samples from people who have a fever. Three weeks later they must once again collect blood from the patients. Each week the helpers must cycle to the Moloundou mission, 25 kilometers from Mbatika, to hand over the filter papers so that they can be analyzed in the lab. The two must be sure not to neglect the *bandas*, because people in these traditional pygmy campsites, many of them deep in the jungle, can also suffer fever.

The next day I hear drops falling on the roof of my workplace. A little later the first Biblical deluge of the new rainy season pours down on the village. Sassi asks village companions to dig a trench around the lab and to reinforce it with banana leaves so as to protect the sides against driven rain. The red soil roads turn to liquid and after a while there is so much earth on my shoes that it feels as if I am walking in clogs. The wooden logs of our luxury toilet are as slippery as soap, and relieving oneself has become a game of skill.

But as suddenly as the rain comes, just as abruptly it stops. Through the mist a man approaches with a snake at least two meters long and as thick as one's fist at its neck. It appears to be a *Bitis gabonica*, or Gaboon viper. Its venom is ordinarily exceptionally toxic, but in this specimen no longer, because its head has been cut off. The adder weighs four kilos. He is too long for my dissection table and so we stretch him out on the ground. I sit on a stool that I move along the serpent by stages so that I can slit it open along its entire length. Sassi looks on, licking his lips. The flesh has the color and texture of cod. After I have taken the required samples, I hand the snake over to the village chef. He chops the snake into four sections and makes as many families happy.

For one part of the team, including myself, it is the last night of our sojourn. We spend it around a gramophone that Sassi has commandeered. It comes with a 45 rpm record, which he has never heard because there is no electricity. With a car battery we bring the apparatus to life.

Sassi cannot get enough of it. He instructs a villager to move the needle back to the start of the groove after every playing. The village chief also dances, like everyone. The women sway their hips and make cooing noises. An old man shuffles backwards and forwards for hours with a bent head and his elbows drawn back, to the same music. By the flicker of the kerosene lamp we drink a toast with our last whiskey. "To a job well done," because there are nearly two hundred organ specimens from 34 different animal species in the liquid nitrogen.

In the morning, the village comes to take its leave of us. We are given three live chickens as presents. These must be taken in the full Peugeot... Some pygmy women who have come to wave us goodbye have pulled truncated pantyhose briefs over their hair. They closely resemble Micky Mouse. How much I have learned to like these people in just one week! I am a little envious of Ethleen, Patricia, and John, who wish to spend a few days deep in the jungle in a banda, in order to understand the manner in which pygmies interact with animals there.

Back in the capital city Yaoundé, providence grants me a gradual return to civilization. I see a very old little woman walk barefoot over the asphalt, bent double under a basket of dead wood. A passing car throws up a great cloud of dust from a hole in the road. The small lady stops to turn away. Then she again walks farther, against a background of advertisements and the signboard of the bar "Chez Irma la Douce." In this 1963 film, policeman Jack Lemmon falls in love with prostitute Shirley MacLaine and tries to rescue her from exploitation. Anti-colonials like me try to do the same with Africa, the alluring continent upon which the rich north shamelessly laid its greedy hands for more than a century. During the past few days I have slept like a rose on my woven pygmy bed in the steaming jungle. Now I cannot sleep in my cool hotel room.

March 1980

Once again I pulverize organ specimens and introduce the ground-up tissue into cell cultures in the hope of discovering Ebola's signature. Again, after weeks of investigation, the result is nil. Ebola isn't in even one of the samples. Laughter deserts me, until I receive a copy of the report that Patricia Webb wrote for the director of the CDC. It gives me pleasure to read that she nominates Dave Heymann and Ethleen Smith for the Nobel Prize, due to their sense of humor while solving practical problems and for creating a strongly bound group from a complex international team. She also prizes my Flemish-English. Hilarious. Only dead sorry that on this occasion, too, we could not contribute one iota to scientific knowledge about the reservoir of Ebola.

FROM EBOLA VIA HANTA TO GRID, FROM GRID TO HIV, AND BACK

How I rise from head of service to head of department, and how I leave the jungle trail of Ebola to venture into the still darker forest of **AIDS** and even find myself with the Bhagwan.

1981

The World Health Organization recognizes the lab of the Institute for Tropical Medicine as a reference and research center for hemorrhagic fever. This is no less than the international approval that we are leaders in this area and ours is an important voice in the house.

We develop new tests to track hemorrhagic fever viruses. Peter Piot confides in me that he does not have the patience for this kind of work and above all else wishes to devote himself to sexually transmitted diseases (STDs). He works closely with the hospital doctors of the ITM, which doesn't just administer vaccinations to globetrotters, but with increasing frequency treats patients from the former colony for tropical diseases such as malaria, yellow fever, or an STD. In the near future something totally unexpected happens in the world of virology that has a greater impact on world health than all the hemorrhagic fever viruses combined. Something that will change Peter's life, and later mine too, decisively.

June–July 1981

The article "Pneumocystis Pneumonia—Los Angeles" appears in the scientific journal *Morbidity and Mortality Weekly Report* from the Center for Disease Control in Atlanta. It concerns five young homosexual men who have been admitted to hospital with PCP (pneumocystis carinii pneumonia), a rare but serious lung affliction. At the time the article appears, two of the five men are already dead.

The newspaper *San Francisco Chronicle* picks up on this article and publishes a piece on lung diseases among gays. Following this, doctors all over the US report similar cases.

Meanwhile my quest for the Ebola reservoir continues in a mood of despair. This time I stay for two weeks in Kinkala, 70 kilometers south of Brazzaville. On the instructions of the WHO, we also investigate the prevalence of the monkeypox virus. But once again we do not find Ebola. Either we are investigating the wrong animal species, or only a tiny percentage of the host species—and there may be more than one—is infected, and we must simply investigate more specimens of each species.

The New York Times reports that the CDC is studying 41 young homosexual men from New York and California in whom Kaposi's sarcoma is present, a cancer of the blood vessels and mucous membranes that until now has mostly been diagnosed in older men in Central Africa. From the investigation it would appear that the immune systems of the young men are afflicted. Through this the patients contract "opportunistic infections," which seldom arise in healthy people or give unsolvable problems, but in people with a bad defense* they are much worse. So it is that patients on the American West Coast develop lung inflammation, serious gum inflammation, fungal infections, or chronic diarrhea.

December 1981

Every week now in the US, five or six new patients are reported to have Kaposi's sarcoma, a rare lung disease, or another opportunistic infection. At the moment the CDC is analyzing 160 cases. In Great Britain and France this mysterious disease profile also pops up. Because the great majority of the patients are homosexual, the disease is called GRID: Gay Related Immune Disorder.

Early 1982

We begin to suspect that we are not dealing with a typical sickness of Western homosexuals. At the ITM's clinic just as many women as men come through the doors and the patients are from Zaïre. Moreover, doctors in the United States and Great Britain note that the disease is also present in people who inject drugs or have received blood donations, regardless of their sexual orientation and activities.

The French doctor and immunologist Jacques Leibowitch is intrigued by the new plague and launches the hypothesis that GRID is caused by an African retrovirus. According to him, the victims die because the virus neutralizes the body's entire defense system. It looks like our suspicions have been confirmed. Colleague Nathan Clumeck of the Brussels Saint Peter's Hospital lets it be known that all his GRID patients are also Zairean.

The microbiology service of the ITM asks the board of governors for the green light to send us to Africa for an epidemiological investigation. I sit on the board myself as the representative of the scientific personnel and explain in detail why this trip is necessary after all that we have seen in our clinic and in America. But it is Saturday, the chairperson is not the ablest of moderators, we have been sitting at the meeting for many hours, and everyone is tired and longing to go home. I draw a blank. Management does not wish to hold our tight budget to ransom with an expensive expedition that may deliver nothing.

Friday, June 11, 1982

In the journal of the CDC I read that during the last twelve months the organization has received 355 reports of Kaposi's sarcoma and PCP or other serious opportunistic infections. Almost eight of every ten patients are homo- or bisexual men.

July 1982

At a meeting in Washington the acronym AIDS* is proposed for the new disease: Acquired Immune Deficiency Syndrome.

Friday, September 24, 1982

The CDC uses the term AIDS for the first time in a report about three heterosexual hemophiliac patients who died of PCP and other opportunistic infections. There is no indication that the three were infected by homosexual contact or drug needles.

Because there are also reports of infections among Haitians, in the press AIDS is called the disease of the four H's: homosexuals, heroin addicts, hemophiliacs, and Haitians.

Early 1983

As the new year starts, a team of scientists in the French Centre National de la Recherche Scientifique under the leadership of Luc Montagnier is in the thick of it, studying blood samples from AIDS patients. They hope that they can isolate a retrovirus. One of the researchers is Françoise Barré-Sinoussi. Françoise knows that her American colleague Robert Gallo, of the National Cancer Institute, has discovered a human lymphocyte virus that he names HTLV-1. However, Bob Gallo cannot demonstrate that HTLV-1 is the cause of AIDS.

In Luc Montagnier's lab on January 23, Françoise is able to isolate a retrovirus, but does not know whether it is the same as Gallo's HTLV-1. Montagnier contacts Gallo, who immediately supplies his French colleague with antibodies against HTLV-1 and with cells that are infected with that virus. Françoise and her team check the reaction of the "American" HTLV-1 antibodies on the "French" virus.

On February 3, the French team determines that their virus cannot be any form of HTLV-1. They christen it "lymphadenopathy-associated virus" or LAV. A little later Bob Gallo will isolate the same virus from the blood of AIDS patients but gives it the name HTLV-3. Today we name this virus the Human Immunodeficiency Virus type 1, or HIV-1, and this is the virus that causes AIDS.

Thus various scientists contributed to the discovery of HIV. Nonetheless, they later argued ferociously about this. At least until 1993, bitter arguments rage with Robert Gallo, who said in 1984 that it was not Montagnier but he who had discovered the virus. Accusations are flung back and forth and an external commission is appointed just to settle the issue. The compromise is that while on the one hand Luc Montagnier's group may be called the discoverers of HIV because they were the first to isolate the virus, on the other hand Bob Gallo's team was first to discover that HIV is the cause of AIDS. When Montagnier's group first published its discovery, the cause of AIDS "remains to be determined." It has since become generally accepted that Gallo's team delivered a great number of the scientific insights that enabled the virus to be discovered, such as a technique to culture T-cells. In any case, in the year 2015, HIV still circulates and, although we have in the meantime found out how to fight it effectively, it is still claiming victims and we still do not have a vaccine.

Spring and Summer 1983
As soon as Peter Piot hears the news about the cause of AIDS, he is more determined than ever to go to Zaïre and show that HIV is present there. But our governing council has said "no," and thus Peter decides to seek funding elsewhere. He meets Thomas Quinn and Richard Krause of the American National Institute of Allergy and Infectious Diseases (NIAID*). Both gentlemen have just returned from Haiti, where they have learned that many Haitians worked in Zaïre during the 1960s. Thomas, Richard, and Peter are convinced by this that there must be a link between HIV and Zaïre. Joe McCormick of the CDC is also pondering this possibility.

The American biochemist Kary Mullis invents the PCR* technique. Ten years later he will receive the Nobel Prize for this. Polymerase Chain Reaction (PCR) is a technique that allows one to reproduce even a minimal amount of DNA rapidly and in an unlimited quantity. An important application of PCR is the genetic fingerprint that allows criminals to be identified. But with PCR you can also hunt down biological felons, faster and with greater accuracy than by antibody tests or by culturing a virus. With it you can elucidate the genetic code of viruses and bacteria, without the need to culture them. Because of this, the hepatitis-C virus was never cultured, but identified and described by means of PCR.

October 1983

The World Health Organization holds the first great conference on the AIDS problem.

Peter Piot, Joe McCormick, and Thomas Quinn travel to Kinshasa and thereby lay the foundation of "Project SIDA." SIDA is the French acronym for AIDS. The project will kick-off in June 1984, with the CDC and NIAID as its most important sponsors, donating 2.5 million and 1 million dollars respectively. Where Antwerp retreats Atlanta dares, albeit this is a good year later and the Americans have a lot more money. Ultimately, the Institute of Tropical Medicine finds a sponsor who contributes half a million dollars.

The Zairean Minister of Public Health supports Project SIDA and puts Peter in contact with Dr. Kapita of the Mama Yemo Hospital in Kinshasa. From this a large scale Zairean-American-Belgian research program evolves that will come to be seen as extremely successful. The program will answer questions such as: how many people are infected, who has AIDS, and what are the risk factors? How is AIDS transmitted and how can we prevent and hinder the disease? The results of the program are published in *The Lancet*, the most important clinical journal in the world.

I am now the head of the Division of Virology within the Microbiological Service of the ITM. Karl Johnson and C.J. Peeters send me an invitation for a working visit to Fort Detrick, a research facility of the American army. Fort Detrick began to work with viruses in a maximum security lab even earlier than the CDC. Fort Detrick was the center for America's biological warfare program from 1943 until 1969. After this was stopped, the researchers applied themselves to biological defense. I do not know if that makes much difference, practically. Just as I have done for my Russian colleagues, I travel to the United States to lend a hand with Ebola research.

My trip to Maryland teaches me just how important serendipity* can be. While I am handling an Ebola sample in the maximum security lab at Fort Detrick, a request arrives from Professor Desmyter of Leuven in Belgium. He asks if Fort Detrick can analyze clinical specimens from lab personnel who have worked with rodents and who by this means have possibly been infected by hantavirus. At this time no one in Belgium can identify hantavirus in a sample. For the Americans it is wholly logical that "the Belge" should busy himself with this Belgian request. I do not know hantavirus, but my colleagues quickly show me the ropes and the relevant test procedures. The virus is named after the Hantaan, a river in South Korea. Hanta is spread by rodents and causes hemorrhagic fever and kidney failure. Thus in hunting Ebola, I encounter hanta. I take the know-how of the virus tests* back to Antwerp and, in cooperation with Professor Desmyter and other scientists, I start an epidemiological in- vestigation of hantaviruses in Belgium.

Army doctor and kidney specialist Jan Clement immediately bites into the subject, due to the link with soldiers who became infected in the rat-infested trenches of the Korean War, and due to the kidney problems that the virus causes. He will build a team at the Rega Institute of the Catholic University of Leuven to study outbreaks of nephropatia epidemica (NE, epidemic kidney disease) in Belgium. NE is caused by the hantavirus pumala. In 1985 Jan establishes a hanta reference center. He has since diagnosed 2,200 cases of NE. The outbreaks do not occur only in Belgium, but also in neighboring countries. The number of new cases of NE in Belgium is increasing, which might be connected to climate change.

1984

Project SIDA isn't out of its starting blocks when I attend the first AIDS conference in the South African city of Johannesburg. I hear speakers state categorically that HIV is only present in undisciplined white homosexuals. During the interval I become acquainted with Miss Mogobe, a nurse. I remark that there are few black or mixed race colleagues sitting in the hall. Miss Mogobe replies that as a black person it took all her powers of persuasion to gain access to the conference. She works in the shanty towns of Soweto, she relates, and has known for some time that not only white men are infected with HIV. "I want to invite everyone in the hall to Soweto in order to see with their own eyes that AIDS has already been raging there for much longer, also in blacks, and also in women," she argues with fire. She doesn't need to convince me.

"If you can support what you are telling me, you have earned yourself a place behind the lectern," I encourage her.

The next day Miss Mogobe is nowhere to be found. During the closing session, after the concluding remarks, the chairman asks if anyone would like to add something. I raise my hand, congratulate the organizers on their initiative, and then throw in that we at the ITM in Antwerp treat both women and men who come from Zaïre and who have precisely the same symptoms as the American homosexuals. I explain that yesterday I spoke with a nurse from Soweto and that she too sees men and women with AIDS symptoms every day. "Mister Chairman," I continue bravely, "I think that it is worth the trouble to check if this lady is right."

A short silence follows, after which the chairman without a blink asks: "Any more questions?"

Meanwhile, HIV is striking the world so mercilessly that within the ITM ever more voices clamor to make this virus a priority. Long meetings follow, in which Peter Piot and I are not always on the same wavelength. If we apply ourselves to HIV, it would be better to demolish our maximum security lab and use the space for HIV research. It comes down to the fact that the expertise we have acquired with Ebola, and since then with hanta too, must make way for the AIDS virus. Again I stand before an intrusive redirection of my scientific career, although my transition from bacteria to viruses was a lot easier than this. The insight that AIDS could become an immense global problem forces me to cut the knot.

Shortly thereafter I receive the instruction to go to Paris and learn how to isolate HIV from patients in the lab of Luc Montagnier. This appears to be somewhat more complex than for Ebola. It took a very long time before scientists had the technique down to a fine art. You need to extract lymphocytes from people. These are the perfect host cells in which to grow HIV—at least if you can keep the cells alive for long enough. During my sojourn with Montagnier I will learn how to do this.

I spend long and intensive days in the lab, which is located in a student district. After work I like to wander around there, sometimes until late evening. Even then light still shines from the many galleries in which budding artists display their works. I enter them and make notes on the little cards that I always have in my breast pocket. The exhibitors take me to be a Belgian journalist. I leave them in the dark because in this way I meet all sorts of young, creative people.

Back in Antwerp I train Greet Beelaert and Gaby Vercauteren, one of my PhD students, in the technique used to isolate HIV. The renovated ITM lab must be ready in 1985 for AIDS research.

1985

In the US, the Food and Drug Administration approves the ELISA test, an HIV diagnostic kit developed by Bob Gallo of the National Cancer Institute. The ELISA is the first HIV test to come on the market. It is crucial to be able to screen large groups of people quickly, especially in blood transfusion centers. This is the only way to know if someone is infected with HIV, because on average it takes ten years from infection for the clinical symptoms to appear. The French Institut Pasteur does not want to use the American test due to a judicial dispute over its patent. Eventually the French and Americans come to an accord, but in the meantime untested blood is given to French patients. In this way more than nine hundred Frenchmen are infected by HIV during the course of 1985, of whom three hundred will not survive.

1985–1986

For almost seven years we have been able to study deadly hemorrhagic fever viruses in the world's smallest maximum security lab. With the arrival of HIV we are obliged to adapt our MSL. The safety rules for Ebola are not essential for work with HIV. The closed isolator can make way for a so-called laminar flow cabinet, a half open space within which sterile air continually flows from top to bottom, is sucked out, and filtered. If I should screw up with HIV, then it can't fly around the lab and instead stays inside the safety cabinet. The vertical airflow prevents the contamination of our cell cultures with bacteria or molds. From now on in our renovated lab we grow and handle HIV viruses that we isolate from infected patients. We also develop new techniques to diagnose retroviral infections.

As soon as the first HIV antibody test is for sale, we use it in the ITM. With the ELISA test we can confirm an infection and start appropriate treatment early, and thus we can raise the chance that patients will enjoy a high quality of life. The Belgian government makes the correct decision to collect epidemiological information about AIDS as quickly and as systematically as possible. For this purpose, a limited number of labs that have the know-how to undertake precise tests are set up. Thus seven AIDS reference laboratories (ARL) are recognized in Belgium. Our lab is one of them. At first it is led by Peter Piot, and after his departure for Africa, by me. It is our task to gather epidemiological information about AIDS in Belgium and to improve diagnosis and treatment of the disease. Very soon we have three new infections per day in Belgium. In 2015 this is still the case.

1986

Was HIV already present in 1976 in Yambuku? We put this question to the Virology service. We retrieve the Ebola samples that we harvested out of our deep freezers. Of the 695 sera that we screen, 5 appear to be HIV-positive! This is just 0.8%, but it is indeed proof that the AIDS virus was already circulating in the African jungle in 1976, seven years before Luc Montagnier isolated it in a Paris lab.

I have by now led the Virology Division for almost three years, and I am now also responsible for the AIDS reference center. I must leave the lab work largely to my colleagues, because my time and energy is spent on management, lobbying, writing publications, securing funds... The latter, especially, burns up time. Only one third of the ITM budget comes from the government. For the remaining cents we must break down doors. Not my thing. I prefer to entertain the dinner-circuits and parish halls of Flanders with talks on AIDS. Today the Progressive Socialist Women, tomorrow a liberal hobby club. The scenario is usually the same. The organizer hands me a cold pint, the sound system is tested, the introduction is read, and the audience casts inquisitive glances at the gentleman who has come to speak about the strange disease. And then after my talk, always the question of whether HIV can be transmitted by mosquitoes. "Well," I say, "there is not a single indication of this, but of course you are entirely free to use a mosquito net. Search under that net thoroughly for mosquitoes 1.6 to 2 meters long, because these specimens can infect you with just one jab!"

1987

The STD-AIDS clinic of the ITM is transferred to the Antwerp Academic Hospital, where Bob Colebunders takes on the task of leading it. Bob is one of the first doctors to work on Project SIDA. He writes an article in connection with this that definitively places AIDS in Africa on the map. It is groundbreaking, because most people still regard it as a disease of Western homosexuals.

To better inform the public and risk groups, the ITM supports the establishment of AIDS organizations such as the IPAC-AIDS Coordination Center and Aidsteam. Already in 1985 we have launched Stag/Aidstelefoon and in 1990 De Witte Raven (The White Raven) will follow. For years I am the manager of Aidsteam. It is not enough to diagnose and treat, scientists must also work in prevention.

Meanwhile, we continue to conduct research aimed at delivering better tests. Bob De Leys of the Belgian firm Innogenetics comes to see me and relates how the new company wishes to apply itself to HIV diagnostics. In the first instance the firm can use the lab of the ITM. After a short while that becomes too small, so I act as an intermediary with the province. Under the leadership of CEO's Hugo van Heuverzwijn and Eric Tambuyzer, Innogenetics can jump out of the starting blocks definitively on the eleventh floor of the provincial building next to the ITM. Later I will also cooperate regularly with companies such as Tibotec and Virco. It doesn't hurt me that some other scientists see something wrong in cooperating with businesses. If science in general and my service in particular improves because of it, then you must shake off this kind of commercial cold water fear.

Friday, March 20, 1987

The American FDA approves azidothymidine (AZT) as a treatment for HIV and AIDS. Only 25 months pass between the moment that it is shown in the lab that AZT can delay the evolution of the disease and the moment that the drug comes on the market. That is exceptionally quick. But all too soon it appears that HIV develops resistance* to AZT.

Saturday, December 26, 1987

Bill Close visits Belgium, together with his world famous daughter Glenn—indeed, the actress who in the same year is starring alongside

Michael Douglas in *Fatal Attraction*. World citizen Bill was born in the US but grew up in France and was an army pilot during World War II. After his training as a doctor and surgeon he left for the Congo. This was 1960, the year of independence. He would stay there for sixteen years, go through the Zairisation era, and become director of the Mama Yemo Hospital in Kinshasa. It was then that he was even appointed as the personal doctor of Mobutu.

By now Bill has been back in the land of his birth for ten years. He runs a general practice in the little village of Big Piney, in the state Wyoming. Dr. Bill wishes to serve his patients as best he can, and in contrast with the American norm, he undertakes house visits. Yet as a scientist he remains fascinated by what took place in Yambuku. For that matter, Bill had a significant share in the attack on the epidemic. Thanks to his direct line to the Zairean president, the Americans from the CDC, well-funded, could come to offer help quicker than expected. Also aircraft and helicopters could be arranged more easily for transport to remote Yambuku. Bill also fixed it so that Dr. Margaretha Isaacson could fly from Johannesburg to Kinshasa in the shortest time with Marburg sera in her baggage, despite the troubled relations between Zaïre and South Africa's apartheid regime. Furthermore, he organized protective clothing and facilitated all manner of other services.

Bill and I had regular contact in 1976. He loaned me his camera. For him, too, it was a trying time, about which he often talks with his daughter Glenn. And so the idea to write a book about the epidemic in Yambuku ripened, a novel to be more precise. Just like her father, Glenn fell under the spell of Ebola and now they are traveling together through Europe and Zaïre to interview witnesses of the drama. Glenn is also thinking about a film and has even enlisted a script writer. Bill wants to finish his novel before Glenn starts with her film, and I sense a slight tension between father and daughter when I accompany them to the convent of 's Gravenwezel. There we will visit Sister Genoveva and her colleagues. Geno was my announcer at Yambuku Broadcasting Systems, the improvised "TV station" in Yambuku that screened reports, which I had filmed with Bill's camera. Now the frivolous but brief venture of Sister Genoveva into the world of film and the media will have a completely unexpected sequel, when the nun stands eye to eye with one of Hollywood's greatest actresses. I see that the meeting with Glenn Close does not leave Sister

Genoveva and the other sisters unmoved. Moreover, Glenn appears to be pleasant company. Also for me. When we part I receive a most agreeable kiss from the actress, under the gaze of the chuckling nuns. I will soon be seeing her father again, because Bill plans to tour through Europe next summer, this time with his grandson.

Saturday, January 9, 1988
Two weeks later I experience a new trip down memory lane, when Karl Johnson of the CDC comes to lodge at my home in Kontich. It doesn't happen every day that you put a hemorrhagic fever legend like Papa Machupo to bed! Karl remains an intriguing figure and a man with brilliant ideas. We talk and drink long into the night and even the perils of marriage slip from the tongue. He and Patricia are no longer a couple. Our discussions wander freely, far from our mutual professional interests. So we both declare our love of sculpture during a visit to Middelheim, the open air museum in the Nachtegalenpark south of Antwerp.

Beginning of 1988
Start of the Human Genome Project,* an overly ambitious scientific program to map the entire structure of the human genetic code by identifying and localizing all human genes.

June 1988
Bill Close and Keir, his 15 year-old grandson, are in Antwerp. Ilse, my oldest daughter, guides Keir through the city and I do the same with Bill. It is a pity that the summer has begun with drizzle and it is between showers that we wander in the shadow of the Gothic Onze-Lieve-Vrouw cathedral and along Brabo to the sixteenth century town hall. Bill is impressed by so much old culture. "You civilized people still have something to teach America," he admits. Indeed, Antwerp could not be more of a contrast with Big Piney, a little village of a few hundred souls on a remote plain set against the backdrop of the rugged Rocky Mountains. According to his own words, Bill feels that he is in his element there, after all his meanderings.

On the previous occasion, when Glenn was with him, Bill had promised the sisters of 's Gravenwezel that he would return once more. This we do. Again we are received with a smile by Sisters Mariette and Genoveva. Only this time Sister Marcella sits there, in a daze. Struck by cancer, she is just a shadow of the enterprising woman in a white outfit who I remember from my Yambuku days.

Bill and Keir have already been away for a while when at the end of June I receive the first draft of his novel. So he is serious. It is accompanied by a short letter in which Bill asks me to read the copy critically and to give the characters names that will be acceptable to the average Flemish reader. He also wants his book to be published in Dutch. A person who asks me such a thing must be conscientious, and I set down my observations in detail on paper. I even doubt if he indeed feels so much "in his element" as he claims, there in Wyoming. It wouldn't surprise me if the country doctor would like nothing more than to break his isolation.

In 1990, when I stay for a few days on his ranch in Big Piney, I establish that he has a Telex in his basement that almost uninterruptedly spews out reports from Zaïre. It appears that Bill still maintains close contact with people all over that country. "I have always known that you are a CIA communications officer," I joke. Bill neither confirms nor denies. Do I spot the trace of a roguish grin on his lips, or is that just my imagination?

September 1988

I fly to Indonesia, on the trail of AIDS. I will be appreciating the comfort of business class more often, especially when I am once again swooping between continents. And always I take along some good reading, often a literary newspaper such as the *Nieuw Wereldtijdschrift* (New World Journal). The poems are so beautiful. At the athenaeum I would have nothing to do with poetry, but at present I cannot get enough of it. This time I travel in the company of Jean-Paul Sartre, Pou-Poul to his friends, and Annie Cohen, his charming biographer. The book is thick, but fascinating.

December 1988

From now on December 1 is World AIDS Day, a day to make the planet's population aware of this disease and to confront discrimination against people with HIV. Shortly thereafter I leave for India on behalf of the WHO to take stock of all that they have done against HIV and AIDS in this immense land. I have much to research and immediately learn that Indians sometimes shake "no" when they mean "yes."

Experienced African travelers who pretend to know what culture shock is can spare their scare stories until they have seen India. We ready ourselves to fly from Bombay to Poona, but the domestic flight is canceled and thus we must cover the distance by taxi. It is at most 150 kilometers, but the trip becomes a nightmare. For many kilometers after leaving the big city I see hovels that have been erected with sticks and rags. The appearance of scrap metal, waste wood, plastic, and other junk between the miserable little dwellings is disconcerting. The stench also—a fetid mix of smoke and exhaust gases.

Poona lies at a high altitude and thus the road winds through steep mountain passes. This is a relief, but the taxi must stop frequently to let the engine cool. Moreover, the gearbox is faltering. I am alert to all that is happening—and hopefully the driver too—except when a herd of cows suddenly crosses the road, right next to a hairpin bend. The beasts are holy and thus the taxi driver slams on the brakes. The wreck begins to skid, I tense all my muscles, my heart stops... and happily so does the taxi. As the dust settles, I see that we are right on the lip of a ravine. Not one meter remains. I gradually become aware of the lowing and slow trampling of the hooves of the phlegmatic cows. While I catch my breath, the driver silently adapts himself to the situation. He maneuvers the car in the right direction between the cliff and the cattle and off we go.

In this former British colony you drive on the left, but the rule is often tossed to the wind. I haven't yet come to terms with what could have been a certain death, when we avoid a head-on collision by a whisker. My despair would be complete if it were not for the appearance of a beautiful panorama now and then. The region reminds me of the Rift Valley in Kenya. But I do not experience real enjoyment, and when we arrive at the Blue Diamond Hotel a few hours later, I immediately crawl into bed, ill from tension and exhaustion.

When I awake the next day, I have stomach cramps and diarrhea but an extensive visit to the National Institute for Virology is scheduled. The day turns out to be instructive and, fortunately, my innards settle peacefully.

Still in suit and tie we want to visit the new *ashram* of Bhagwan Sri Rajneesh in the late afternoon. Coincidentally, the fabled guru has created his own "district" just around the corner from our hotel, and of course we are curious as to what may be experienced there. We walk past countless little shops and three wheeler taxis that slalom skillfully through the traffic. There is a conspicuous number of whites in this neighborhood. Signboards hang at the entrance to the ashram: "This village is AIDS free" and "No pictures." At the reception four ladies are standing in jeans and polo-necked blouses. We feel terribly overdressed and after some discussion agree to come back the next day. I leaf through the guest book and see many Dutch names. I also note that people in this sanctuary openly caress each other. Intriguing.

The following day Amrito and Indivar await us at the entrance to the center. They are both doctors and of American origin. Before we are taken to the headquarters, I am frisked by a young woman. "You are the doctor from the world health club," she says, while she remains looking at me for a couple of seconds. Odd that both physical and eye contact last meaningfully longer than back home. Club? That's hilarious, I think.

We aren't carrying any weapons and that is rewarded with coffee and cookies. Thereafter we sit in a spacious cinema to view a propaganda film in which Bhagwan addresses themes such as the hole in the ozone layer, Hitler, and extremist Jews. His pet subjects contrast wildly with the green décor with its lovely laughing faces, flowers, and little birds. But the climax is still to come: the evening meditation under the leadership of the Bhagwan himself. Indivar invites me to attend this, but there is one condition: I must prove that I am not infected with HIV. That makes me think about what it is that is exchanged between enlightened souls. At once I understand the little sign at the entrance. Also with regard to taking photos. In any case, my word of honor does not suffice and I must allow blood to be drawn. Meanwhile it is already 18:00, and I am expected by Amrito and Indivar at 18:30.

On with it, then, although I demand that the nurse uses a sterile needle before she pokes one into my arm. While my blood is being tapped I draw the attention of the health care workers to a few mistakes that they are making. Yes, when people force me to obey stupid rules, I become peevish. We're finished. What now? I know all too well that no one in India has an HIV-test that yields a result within an hour. My blood sample will be sent to one of the two private clinics in Poona. Happily, while waiting for the result I am allowed to enter the meditation room. I learn later that in the case of a positive result it will be sent to the reference laboratory. In the past year nearly 4,000 Bhagwan fans have been tested according to this procedure, of which three appeared to be seropositive.* What a nonsensical safety procedure. And talking of safety, my test result will at the earliest only be known tomorrow. Thus I can still romp around in the ashram for a full night putting others at risk.

At about 19:00 the tension rises. I am asked to take off my sport shoes before I, as a special guest of Indivar, may join the long queue of waiting people. Everyone has a ticket, because the disciples must pay to hear the words of the master. We shuffle into a connecting room, a kind of circus tent with a hemispherical blue ceiling. The sides consist of mosquito nets. Electronic music plays in the background. We sit on small cushions. Some pamper themselves by settling on luxury versions with backrests—not a bad idea, perhaps, if we must sit and listen for two to three hours. I introduce myself to my neighbors, an old American woman of Indian origin in a wheelchair, flanked by her son. He is a surgeon.

The hall is full. The tempo of the music quickens. People close their eyes and move with the rhythm, letting themselves go. Some call and whoop ceaselessly, until the music stops and a collective scream resounds, as if from one mouth. Arms are lifted into the air, fingers spread wide. This ritual is repeated a couple of times, after which utter silence intrudes and everyone awaits the coming of the Bhagwan, immobile and with staring or shut eyes. A sultry woman's voice announces the Bhagwan's arrival. Outside, the tires of what must be one of his famous Rolls Royces crunch on the gravel. Then here he is. He strides in. The Bhagwan is wearing a somewhat droll black cap, simple dark attire, and sunglasses. When he folds his hands in front of his face, his long white beard becomes more conspicuous. He greets the public very calmly, and in complete slow motion sits in a gigantic white chair and begins to talk, in reasonably good English.

Very quickly my buttocks become numb. I appear to be the only person who must continually change position. Fortunately I later spot several other people who now and then shift a leg or a foot, although they do not do it as visibly as I feel I am. The surrounding believers listen as statues and with closed eyes to the words of their master. The loudspeakers issue not only his voice but also the sounds of a tropical rain forest. Everything is recorded, so that the disciples and those who are absent can listen to the speech with a cassette player, for a price, of course.

The Bhagwan says things such as: "The master has nothing else to do than to lead you to self-realization. You cannot exercise relaxing. A relaxed feeling has to do with your entire inner being. Only in America can you learn from books how to relax." Or again: "All traditional religions are written for death. Life demands openness. The more you open yourself, the more you will be able to live a real life."

After two hours of meditative tittle-tattle he throws a few burning chestnuts at his listeners, maybe to sharpen their sexual appetite. Then follows a rumble on the drums and everyone falls backwards. I do the same and land against the wheelchair of the American-Indian grandmother next to me. My front neighbor in his turn lands on me. I ignore this and remain for about ten minutes lying in this position, while listening to the epilogue of the Bhagwan.

A new attack on the drums and the public once again sits upright. The rhythm of the music becomes frenetic and the bodies around me begin to twist and turn like caterwauling cats. They make just as much noise. I take part in it all while trying to remain outside of it. Finally the Bhagwan rises, just as slowly as he sat down. Several hysterical women accompany him to his English sleigh. In the hall people linger for a while afterwards, but I follow Indivar. He invites me to a meal at the Bhagwan Club. Outside, in a wonderful temperature and by Bhagwan candlelight, I enjoy a Bhagwan pizza with cool Bhagwan water.

However unusual my visit to the ashram may have been, I am not the first of my relatives who has been involved with the Bhagwan. For a year Dora van der Groen walked around in the red clothes. "For me it was a means to learn how to live in the here-and-now. That was good for me. Now, that was in the years 1979–80," the actress disclosed in *Dora van der Groen* by Erwin Jans and Klaas Tindermans.

Four days before Christmas I arrive at Zaventem after hours of delay in New Delhi and a missed connection in Frankfurt. The cold, the clouds, the Christmas decorations—everything in my country shouts winter. It gives me the blues. Always on the road. Much time spent to learn little. But what is better: to spend days and days traveling to experience unpredictable moments intensely as they present themselves? Or to bury yourself deeply in the best articles and books about a country? There you have the brass tacks. But no text can describe the appearance of a ministry in New Delhi with sufficient force. You may well seek artful words to say that such a building is dark and worn, that a fan spins every three meters and that against all reason the officials are jovial and lively. But the lethargy that hangs in the passages, the melancholy that clings to the dilapidated walls, the lack of will to improve the world or at least to be proud of one's own work… These things you must breathe in, taste—gagging a bit as you do so, and afterwards, shivering, shake out of your bones, like a quack who lays his hand upon a patient to draw off bad energy. A book can capture your imagination, but travel tattoos your soul. No Sartre can best that. Perhaps a combination is best—reading while traveling—and that I do. This thought lifts me. I have been at home again for several weeks when a friendly letter from Poona drops into my letterbox with the news that I do not have HIV.

Spring 1989

I must give a few talks in Israel, and Dina can come along. In the Tel Aviv Hilton I give four lectures to people who only speak Hebrew. I talk about the fast HIV antibody tests and AIDS in Africa. Meanwhile, Dina travels in a luxurious Mercedes with a chauffeur to the Crusader town of Zikhron Ya'aqov. In the evening we wander beside the Mediterranean Sea. There is no better time than this to visit Tel Aviv, Hebrew for "spring hill."

My professional transition to AIDS doesn't occur overnight, of course. Currently my colleagues and I still publish a few more scientific articles about Ebola, hantavirus, and hemorrhagic fever than about HIV and AIDS.

I write some articles in cooperation with my Russian colleague Dr. Rytik, who visits Belgium. Over the years we have become friends. Attentive as always, he has brought small gifts: matrushka dolls, a bottle of Zubrowka vodka—the one that has a bison on its label—and a double LP with profoundly beautiful Russian Orthodox music. I have traveled so often to the Soviet Union that I have become extremely fond of Russian music and art.

I recall in sharp detail a visit to the famed Tretyakov Gallery in Moscow. This museum houses the greatest collection of communist propaganda and portraiture. In contrast to my Russian colleagues, I rush like Sputnik through the halls where the red flag waves from the paintings to a point farther on where I am struck like Paul on his horse on the road to Damascus by the aspect of an immense nineteenth century canvas that was painted by one Ivan Aivazovski. I do not know this artist, but he appears to be one of the best marine artists in history. In any case, the seascape that he has conjured nails me to the spot here and literally takes my breath away. For minutes I stand and gaze at it. Aivazovski, a fantastic discovery.

Two months after Tel Aviv I am once again underway. Airports have become my second home. I stopover in Anchorage International Airport, Alaska. This gives me a chance to sketch these business class lounges where you can sip away your jetlag in comfort, drink in hand, snuggled into an armchair on a thick beige carpet, with muted light and softened music. A glance at the world map tells me that I have once again taken the longest flight route in the world: via Anchorage over Tokyo to Seoul. When I arrived here it was 20:45. From here I go to Japan, a flight of 6.5 hours. On Sunday I arrive in South Korea at 09:00 Belgian time and 16:00 local time. I have been travelling for 27 hours.

I find it difficult to climb out of bed in the morning during my first few days in Seoul, because my body clock still stands at midnight. I select several classic WHO sessions and see a couple of familiar faces, such as that of Karl Johnson.

October 1989

In the journal AIDS, Martine Peeters, one of my previous doctoral students, publishes an article in which proof is given that in wild chimpanzees in the West African country of Gabon, a virus is circulating that closely resembles HIV-1. Big news, because the question immediately arises as to whether this chimpanzee virus is perhaps the ancestor of human HIV. An investigation is started in which 44 chimpanzees, the majority in the Antwerp Zoo, are checked for antibodies against HIV and SIV, the ape variant. In one chimpanzee, Noah, antibodies against HIV-1 are found and a virus which is related to HIV is isolated, but it differs from the HIV-like virus that Martine discovered in the wild chimpanzees of Gabon. Noah is a sturdy lad, about five years old, who was intercepted by Belgian customs as illegal cargo from Zaïre a few years ago.

The Zoo's managers would rather die than have their clientele associate the Antwerp chimpanzees with AIDS. The business must remain behind closed doors and Noah, together with an uninfected friend, is transferred to the Biomedical Primate Research Center in the Dutch town of Rijswijk. The BPRC is the largest primate research center in Europe and above all else conducts research into diseases that threaten human health. In Rijswijk, Noah will be followed for many years, for example by Pascale Ondoa, another doctoral student in my department.

Late 1989–Early 1990

Today, in Isaac Newton Square in the town of Reston, Virginia, 35 kilometers from the American capital Washington DC, there is a children's day care center. In late 1989 on this site Hazelton Laboratories set up a center to do animal experiments. Initially some 500 macaques were brought here and in September roughly another 100 animals were added. These are Java monkeys from the Philippines, which have been flown in via Tokyo and Amsterdam to New York, and then shipped from there to Reston in a truck.

Before the experiments can begin, the new animals must pass quarantine—a consequence of the deadly incident with infected monkeys in Marburg. During the isolation period, 29 of the 100 Java monkeys in Reston die. Hazelton Laboratories faces a riddle. During the autopsy of one of the animals the veterinarian discovers that its spleen is three times the normal size and there is blood in the intestines. He dissects a few more of the monkeys and comes to the conclusion that they all died from a SHFV infection, a virus that causes hemorrhagic fever in monkeys.

In order to confirm his finding, the veterinarian sends monkey tissue samples to the United States Army Medical Research Institute of Infectious Diseases (USAMRIID), a military establishment. Anticipating the result, workers begin to euthanize the remaining Java monkeys. In the meantime, monkeys die here and there in other rooms in the center. On November 28, a USAMRIID researcher who views the tissue samples under an electron microscope discovers that it isn't SHFV but the Zairean Ebola virus that is responsible for the outbreak. The report is extremely unsettling because since 1976 all virologists have known that when Ebola is transmitted to people, in most cases it means certain death. Because monkeys which had no contact with the Filipino macaques have died, the terrifying suspicion arises that they are up against a new or mutated Ebola variant, one that can be transmitted through the air. This is the ultimate horror scenario: a "supervirus," just as lethal as Ebola and just as contagious as the common cold.

In this case the contagion would have spread via the air-conditioning of the building in Reston. USAMRIID doesn't let the grass grow under its feet, because if the Ebola virus escapes from Hazelton's building and reaches Washington DC in short order, the consequences will be inestimable.

President George Bush Sr. is warned. "Ebola is very nearby, a half hour drive from the White House. We must react at once," an adviser informs Bush. Already on the day of the discovery, USAMRIID initiates emergency consultations with the CDC and public health authorities of the state of Virginia. An action plan is immediately devised. The CDC will bend over backwards to prevent people becoming infected with Ebola and US-AMRIID is going to tackle the monkeys—all of them: dead, sick, or healthy.

The army decides that all monkeys in the building at Isaac Newton Square must be immediately destroyed. But that must occur without danger for the people involved—not a simple task in premises where the air is apparently full of Ebola. Scientists are sent to Reston with fully enclosed protective suits. During the euthanasia, one monkey escapes, and some scientists spend the rest of the day hunting it. At last they manage to capture the animal; it was still inside the building. Once all the monkeys are dead, a period of eleven days follows during which the whole facility is scrubbed with bleach and exposed to the vapor of formaldehyde crystals to decontaminate the air.

Eventually the researchers will determine that they are not dealing with Zaïre Ebola but with another, previously unknown variant of Ebola, which is demonstrably transmissible by air but which fortunately isn't deadly for humans. In subsequent outbreaks of Reston-Ebola (presently called RESTV), it will be seen that people who are infected by it do not even fall ill. But that is later.

Meanwhile, Reston has a couple of blood-curdling weeks behind its back. Author Richard Preston found inspiration in these events for his bestseller *The Hot Zone* (1994), a non-fiction thriller. I read the book in one stretch. It sparked imaginations so much that film studios immediately made offers for the rights. The result was the box office success *Outbreak*, a fictional spin-off of Preston's book with Dustin Hoffman and Morgan Freeman in the main roles and a much sadder ending than the book. In the film, a little monkey also escapes, but this one is much more quickly and easily caught—to the great frustration of the scientists involved in the real incident.

The animal experiment center in Reston never resumes its work. The building stands derelict for years until it is demolished in 1995, in order to make way for a children's day care center. Six further RESTV outbreaks in Filipino monkeys will follow, and once, in 2008, also in pigs in that country. The last case was disturbing because pigs transmit diseases to people more frequently than monkeys. However, the six people who were possibly infected by the pigs did not become sick.

1990–1991

In 1990, in collaboration with Innogenetics, the ITM publishes an article in which we show that we have been able to isolate a new variant of HIV-1 from the blood of two people in Central Africa. It is an important finding that proves multiple variants of HIV-1 are circulating. The ELISA test that is being applied on a large scale yields a negative result for an HIV-1-ANT-70 infection. ANT refers to Antwerp. Thus patients with "our" variant are slipping through the holes in the net. For blood transfusion centers this could be a disaster. Together with Innogenetics, we adapt the ELISA test by building in part of HIV-1-ANT-70. My colleague Gaby Vercauteren throws herself into this work heart and soul.

Worldwide ever more HIV antibody tests are developed. So that the forest can still be seen between all the trees, the WHO asks the ITM's Aids Reference Center to act as a kind of test purchasing agent for HIV diagnostics. Gaby Vercauteren joins the WHO as a test specialist, so that henceforth we have a first-class communications channel with this international organization. Gaby processes the results of a study in which a member of our staff, Greet Beelaert, compared more than 36 tests.

In such an investigation two parameters are of great importance: sensitivity and specificity. A test is 100% sensitive if five of 100 people have HIV antibodies and the test detects all five of the positive samples. Sensitivity is good, but a test mustn't be too sensitive, because then the risk exists that you will also find positive results in the 95% of people who don't have HIV. These are so-called false-positive results. The risk of this is real because in a person's blood there are all kinds of antibodies, and if some antibodies that are of no interest bind to the cell material of the vero cells, you have a false-positive result. Therefore a test must also be specific. It must only find what it is intended to find. A test is also 100% specific if in all 95 uninfected samples it yields proper negative results.

To be sure of a positive result, we always do a confirmatory (or verification) test. This takes longer than the first test and is also a lot more expensive. Right from the beginning in Belgium it is understood that in the case of a positive test the patient will only receive his result after the result of the confirmatory test is known. The improvement of HIV tests has always been a chief objective of the Aids Reference Center. We want a test that has maximum precision and reliability. The confirmatory test that we use at the ARC is of the type known as a western blot test. In this highly expensive technique, a specific viral protein is identified in a mixture of proteins by means of antibodies, after they have been sifted on the basis of molecular weight. At the ITM we have refined this principle. The result is a test that we call the Line Immunoassay and which is put on the market by Innogenetics under the name Innolia. I am proud that the ITM cooperates constructively with the commercial world, without selling its scientific soul to the Devil.

And yet we could have played it better. At the end of 2014 I sit with my successor, Kevin Ariën, in the Café Heilig Huisken, the tribal watering hole of the ITM. Kevin tells me that the ITM is going to seek a patent. This has never happened before in the history of the ITM. I have tried to, indeed several times, but the top brass always put a spoke in the wheel. Innogenetics benefitted far more financially from the Line Immunoassay than we did.

I had often asked myself why we couldn't do what the Rega Institute in Leuven did. That center blossomed under the leadership of the erstwhile rector of the Catholic University of Leuven, Piet de Somer. According to him, a university must do fundamental research, but that doesn't exclude the possibility of finding immediate practical applications for the results. Indeed, the Rega Institute applied for patents and in that respect it was very innovative in the Flemish scientific world. Investing in patents and commercializing research results delivers income that can again be pumped into fundamental research. But I could never convince our management of this. I can but applaud the change of course that occurred after my departure.

It may seldom or never appear in the news, but the funding of science is a permanent source of anxiety and frustration. For years I coached a group of people including several brilliant minds, who after five years (sometimes less) would come reluctantly to ask if they would still have work next year. The greatest part of our resources came from project grants and companies. That sort of finance always has an expiration date and within the Microbiology service I must continually fill all the financial pots to keep my colleagues at the mill. In this regard the appointment of nursing expert Jan Vielfont as the organizational and administrative manager of the Microbiology service was a successful strategy. Jan has worked at the ITM since 1975 and dedicated his life to juggling with grant dossiers. He would also initiate many international AIDS conferences, even in Africa.

In Geneva, Tim Berners-Lee, a software developer of the European Center for Nuclear Research (CERN) and his project manager, the Flem Robert Cailliau, conceive the World Wide Web. The purpose is to achieve a smoother exchange of information between the scientists in CERN's diverse international project teams. The importance of their idea can scarcely be overestimated. Internet and the Web will fundamentally change the way people in general, and scientists in particular, interact. A new living space is born in which people, unrestricted and at the speed of light, express themselves, open up new horizons, and propose and record deeds. For science the World Wide Web is a quantum leap without precedent. Researchers who were at most names printed on paper, or voices on the telephone, become recognizable faces, become personally acquainted, inspire each other, and share and comment on their findings. In the fight against HIV, the Web arrives none too early. Up to its birth there have been 6.5 million AIDS deaths and 18 million people who must learn to live with HIV. It is quite possible that the lifesaving vaccine is flying around in cyberspace and it is just a question of time before someone correctly assembles the right pieces of the puzzle.

Sunday, November 24, 1991

Freddie Mercury, the front man of the British rock mega-group Queen, dies. The previous day he let it be known through a press release that he has AIDS. "Following enormous conjecture in the press, I wish to confirm that I have been tested HIV-positive and have AIDS. I felt it correct to keep this information private in order to protect the privacy of those around me. However, the time has now come for my friends and fans around the world to know the truth, and I hope everyone will join with me, my doctors, and all those worldwide in the fight against this terrible disease." More than any other death, that of the rock icon Freddie Mercury rubs the nose of the average citizen of the Western world in the reality of AIDS and HIV.

Other famous Americans who die of AIDS are Hollywood icon Rock Hudson, pop artist Keith Haring, *Psycho* star Anthony Perkins (he acted Norman Bates), the African-American top tennis player Arthur Ashe, pianist and entertainer Liberace, the rapper Eazy-E of Niggaz Wit Attitudes, biochemist and science fiction writer Isaac Asimov, and Tom Fogerty, the bass guitarist of Creedence Clearwater Revival. But also outside the United States the virus fells celebrities such as the Russian ballet star Rudolf Nureyev, the French philosopher Michel Foucault, the Nigerian musician and politician Fela Kuti, and Kongulu Mobutu (son of Zaïre's President Mobutu).

1991–1992

Important changes of the guard. My first boss at the ITM, Stefaan Pattyn, pioneer in the fight against bacteria and viruses, leaves on pension. Dr. Wouter Janssens and engineer Leo Heyndrickx arrive to reinforce the Microbiology service and will undertake molecular biological research on retroviruses. Peter Piot leaves Antwerp and becomes the joint director of the Global Program on AIDS. In 1994 he will be promoted to executive director of the Joint United Nations Program on HIV/AIDS.

August 1992

I am in Washington for the annual AIDS conference that Bob Gallo organizes, and I make use of a free Sunday afternoon to once again visit the Smithsonian Museum. It is my favorite museum, or rather museum-park, because it is actually a cluster of eighteen museums. Here, in no time, you can make crazy leaps between cultures from all over the world. So it is that I travel from the Shang dynasty and a bronze vase that is nearly three thousand years old, via the Museum for African Art, on to Vermeer's *Girl with a Red Hat* in the National Gallery.

A week later Reeboks squeal on the shining floor of the Hyatt Regency Hotel in Washington. Tired congress attendees with name cards walk through the lobby. While I am waiting for a taxi to the airport, I allow my table companions of the past week to pass in front of my mind's eye—people who I will encounter again, often, in some other part of the world—my congress relations. The previous Sunday I ate pizza on my own, but on Monday I went to a Thai restaurant with Jim, Helga, Gunnel, Peter, Eva-Maria, Patricia, and Mark. On Tuesday I go out to eat with Sheila and Deborah,

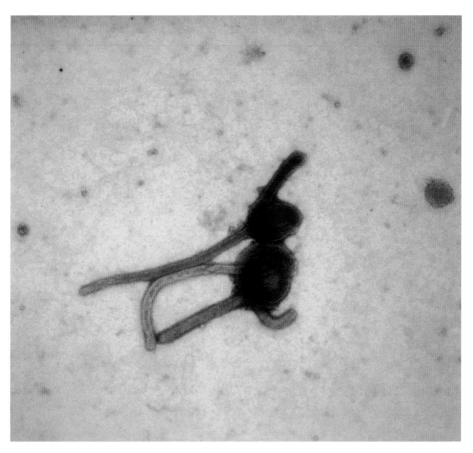

An electron microscope image of the Ebola virus. The virus was first imaged and studied in Antwerp on October 12, 1976.

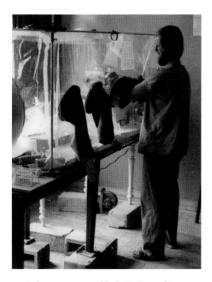

My isolator on concrete blocks in the Ngaliema Hospital, Zaïre. In this construction I can work relatively safely with blood samples and other materials that are (possibly) infected with a deadly virus. (October 1976, Kinshasa)

The fluorescence microscope in the bathroom, the only room I can darken to look at my "little stars." Infected cells to which antibodies against the Ebola virus bind light up when you observe them under ultraviolet light. (October 1976, Kinshasa)

Pictures that my daughters created in faraway Flanders. (October 1976, Kinshasa)

With Father Carlos (left) and Peter Piot (right) on the veranda. (November 1976, Bumba, Zaïre)

The Danish-Dutch-South African Margaretha Isaacson, who focused on the therapeutic aspects during the outbreak. (October 1976, Zaïre)

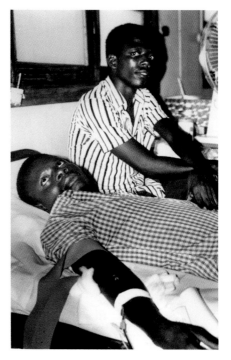

Nurse Sukato Mandzomba (sitting), with fifteen donations, the world champion in plasmapharesis. (November 1976, Kinshasa)

During a day outing on the Dua River I give free reign to my sense of rhythm. (November 28, 1976, Zaïre)

Duo on the Dua with Sister Geno (Genoveva). (November 28, 1976, Zaïre)

Epidemiologist Joel Breman also greatly enjoys himself. (November 28, 1976, Zaïre)

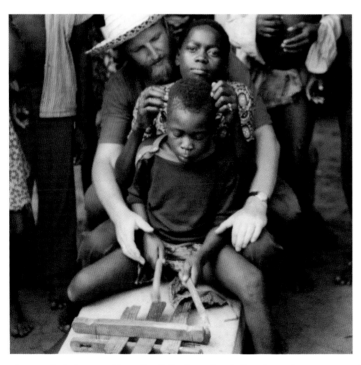

The percussionist in me participates in a six-hander on the xylophone. (November 28, 1976, Zaïre)

Epidemiological field notes with the name of teacher Mabalo, the index case of the Yambuku outbreak. It is the first Ebola outbreak to be investigated scientifically and this will lead to the identification of a previously unknown virus. (November 1976, Yambuku, Zaïre)

From left to right: Sister Mariette, Stefaan Pattyn, myself, Peter Piot, Karl Johnson, Sister Marcella, Mike White, Danny Courtois. Sitting: Joel Breman and a man whose name I cannot recall. (Late 1976, Yambuku)

I pose with a small bottle of IF-conjugate, the magic substance with which we can make Ebola antibodies on infected cells light up under the fluorescence microscope. (November 1976, Yambuku)

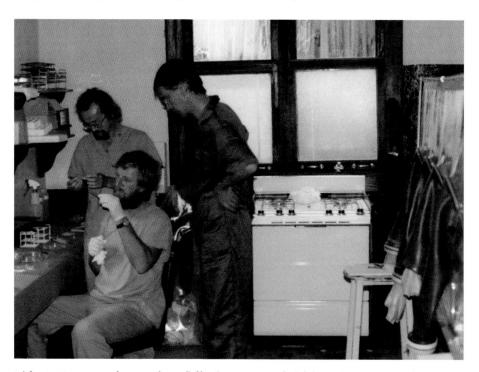

I sit between Peace Corps volunteers Del Conn (left) and Danny Courtois (right). (November 1976, Yambuku)

Skillful Sister Geno knows the people at the mission like no other and always asks them about how things are going at home. (November 1976, Yambuku)

Commission leader Karl Johnson and his secretary Leslie Welch. (Late 1976, Yambuku)

Sister Mariette, the only nun who lived through the Yambuku outbreak and survived. (November 1976, Yambuku)

Sisters Geno (left) and Marcella. (Late 1976, Yambuku)

From left to right: Sister Marcella, Stefaan Pattyn, Sister Geno, Sister Mariette. (Late 1976, Yambuku)

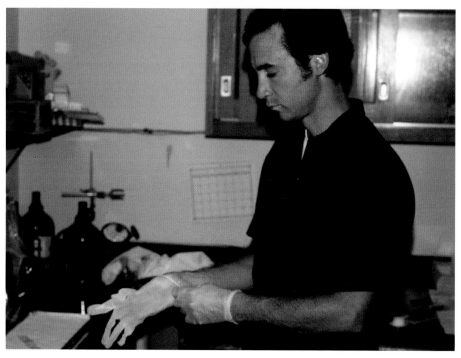

Jim Lange, my training supervisor at the CDC, teaches me the procedures and tricks of handling viruses in a BSL-4 lab, a maximum security lab. (Summer 1977, Atlanta, Georgia, USA)

During my training at the CDC. (Summer 1977, Atlanta)

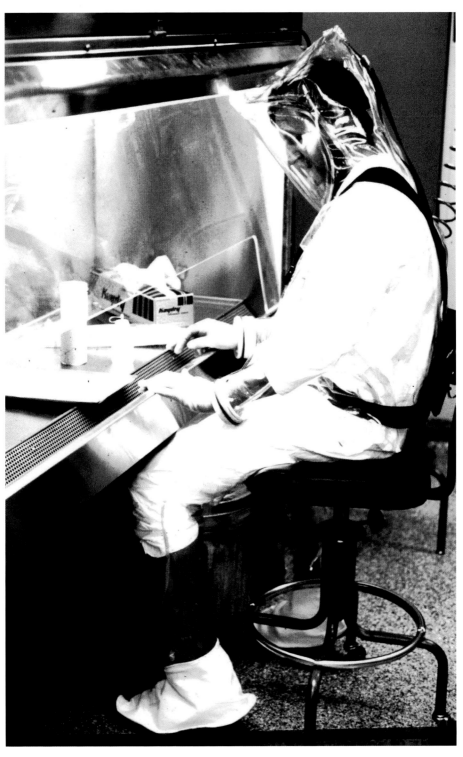

With a pressurized air hose on my back so that no air from the surrounding environment can enter my suit, even if it leaks. (Summer 1977, Atlanta)

Working in a closed isolator allows one to breathe much more easily but hinders movement! (Summer 1977, Atlanta)

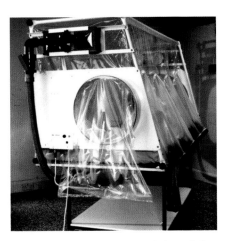

The isolator that I built in the ITM on the basis of what I had learned in Atlanta. (Late 1977, Antwerp)

It is no easy task to move objects in and out of my isolator. I seal an object that I have removed from the isolator inside a plastic bag. (Late 1977, Antwerp)

One of my cartoons, showing that a high security lab is to be highly recommended in the fight against sexually transmitted diseases. (1977, Antwerp)

Arrival in the pygmy village that would become the base from which we would sally forth to catch animals that are possible hosts of the Ebola virus. (February 1980, Mbatika, Cameroon)

I show the Ebola virus to Village Chief Sassi (with hat) and his villagers. (February 1980, Mbatika)

The cadaver of a potential Ebola host on the roof rack of Dave Heymann's old Peugeot. (February 1980, Mbatika)

In the jungle a traditional pygmy banda is set up in no time. (February 1980, Cameroon)

"Our boss is hanging out again," announces my secretary Ciska Maeckelbergh and other colleagues. (1993, ITM, Antwerp)

an American and an Ethiopian. Wednesday sees me with Frank and Fowke, Thursday I dine with Hans and Scott, Friday with Patricia, and yesterday I rounded off the week beautifully with my compatriots Joost, Linda, and Zwi.

After nearly two weeks in Belgium I am back in Washington, to review the state of the game in the development of an AIDS vaccine. The entrance drive to the Westfields International Conference Center is richly decorated with Christmas lights. In August? Am I awake? Have I lost my marbles? When my daughters return to school, after an umpteenth summer vacation during which they almost saw their father, I will not be there to wish them success and to wave goodbye. I just don't know how long I will be willing to do this.

Sometimes I am as happy as a child when encountering the exciting international ambiance that you only find in airports, the tremendous feeling of acceleration as the jet engines push me back into my seat during take-off, the stunning landscapes, the fascinating people who I meet, new colors, and occasionally delicious food.

Other times I just become plain tired of the overbookings, the jetlag, superficial twaddle with other congress stalwarts, and the uninspired architecture of hotels that everywhere in this world look like each other. I am as sick of them as of cold porridge, all the variations on the same insipid theme: the dark marble of chilly lobbies, the stainless steel of the elevators—Otis, Schindler, Kone, ThyssenKrupp, I know them all—and the people inside them who stand uncomfortably in silence and sometimes could do with a shower, the tired stale carpet in the passage that you walk along, the shocking red firehose cabinets, the room key with a giant key tag or the different card systems which demand more than three attempts to open the door and activate the lights, the dark bathroom with protective paper on the WC seat and the end of the toilet paper routinely folded into a point, the suitcase rack with its metal X-legs and wide rubber straps that is always in one's way, the heavy double curtains with exaggerated patterns, the kitsch reproductions on the wall, the little bowl of glazed mints next to the notepaper, the room service that requires orders to be repeated, the seven TV channels of which half have speckled images, the Gideon's Bible in the bedside table, the hard starchy blankets on a bed that has been much too tightly made and in which you must kick like an idiot to create enough space to stretch your legs. And then you must climb out again from that strange luxurious nest because, although you can switch off your reading light from bed, you can't put out the light in the little bathroom passage, and thereafter you stand up yet again because you must use the toilet, and in the bathroom there is that basket with mini soap cakes, shampoo bottles, a shower cap, and even beauty cream—God bless. And after this, the darkness, significantly darker than at home, but disturbed by the intrusive orange glare of the short orange bars that form the digits on the display of the radio-alarm clock and by the sound of the night traffic on the street, two, five, or seventeen floors below. And then you can't sleep. Or perhaps you manage to drop off eventually, but sleep badly until the alarm goes off or the merciless wake-up call comes in rapid, mangled English, after which you freshen up, dress, and hurry down to the breakfast buffet. Following fruit juice and urn coffee, you fetch a dollop of porridgy scrambled eggs from under a copper dome and work everything down mechanically, amid surly-looking hotel guests who have apparently slept worse than yourself, or real early risers who cannot bottle their alien wit and deliver raucous commentary on the much too tender day.

Next year I will have worked for two decades for the ITM. If I am still alive, ITM director Luc Eyckmans will honor me with a red rose. But is it truly healthy to wear down the soles of one's shoes for the same organization for 20 years? Sometimes I get the idea that I have become "Patience on a monument," that above all else I have become a mere name in an address book, a business contact to be used just so long as I fit the agenda. Behind the façade of a successful life lurks a human failure.

Dina, who is approaching fifty just like me, grasps every opportunity to rub my nose in the fact that her life has been ruined by all my years at the ITM. When our oldest daughter marries, Dina does all she can to prevent her landing in the same boat as her mother. Dina has literally said: "Until today the beautiful moments have been very few and far between." The red of the rose that Eyckmans will give me stands for the blood shed by pricking myself upon the thorns, or the blood shed when I pushed thorns under Dina's nails by being at home so seldom and letting her shoulder the burden for everything, absolutely everything.

August 1993
We are one year down the road and I once again sit on an aircraft bound for Washington. Halfway between Europe and the United States I send a

fax to one of my thesis students. Only now have I found the time to read his treatise. These trans-Atlantic flights are made for this. No telephone disturbs me and Dvořák's string quartet Opus 96 plays through my headset; it was created when the Czech composer visited America for the first time. From Dulles Airport I share a taxi to the Hyatt Regency Hotel with three Belgian colleagues. In the lobby I pick up the congress program. Bedtime reading.

Although each year Bob Gallo promises he will limit the program of his next AIDS meeting in both time and number of presentations, this year I am one of the 47 speakers. A Gallo congress is a race. Uninterrupted listening, speaking, and discussing from 08:00 in the morning until 19:00 in the evening. This evening we have a late night session as the cherry on the cake.

After all of this I am addressed in the elevator by a lady from the American newspaper *Newsday*. "Hi, Doctor Vendergroan. My name is Laurie Garrett. I have met two of your friends, Joe McCormick and Peter Piot." Laurie is a Californian biologist who broke off her doctorate to embark on a career in scientific journalism. She would like an interview about Ebola and Yambuku. I agree immediately. Just so long as it's not about AIDS. Seventeen-year-old memories once again rise to the surface. We begin in the hotel's bar and continue in an Italian restaurant, where we are the last customers to leave. Two years later Laurie Garrett publishes *The Coming Plague: Newly Emerging Diseases in a World Out of Balance*. In it she describes how vulnerable our planet is becoming to all sorts of diseases, because we devote too little time and money to health care. The fifth chapter of this studiously referenced book is dedicated to Yambuku 1976 and is partly based on the discussion that Laurie and I had that night.

As always during the course of a Gallo congress, I visit the nearby Smithsonian Museum. First I collect my new membership card. It feels good to be part of the Smithsonian family. I say "Hi" to the staff behind the display counter at the museum shop, where virtually countless books and exotic baubles are for sale. While I wait in the queue, the cashier taps out the rhythm of the African music that plays throughout the day.

The congress is fascinating but exhausting. In the conference room of the Hyatt Regency the air-conditioning is so cold that the outdoor air is like an oven when I leave it to eat at midday. I find it difficult to return to the conference after lunch. For many on the home front it is a well-kept secret that congress delegates often don't see the whole game through. Some skip sessions or simply don't appear on the last days. This conference also comes to an end, and only the diehards are still there. The last talk is given by my Belgian colleague Zwi Berneman. Bob Gallo closes the congress with the words: "I hope that you are taking something away with you." Yes, I think, an aching back and a cold.

That evening Bob throws a farewell party for intimates and old congress hands. It is still warm in the garden. Crickets chirp while we enjoy delicious Italian dishes and wine. In the spacious, cooled salon of the splendid house I chat with Zwi. For five years the young Belgian doctor has worked for Bob Gallo and he reflects with satisfaction upon this time. Gallo has been painted black by the row with the French Institut Pasteur over the commercial rights to the ELISA test and the deaths caused by it. Bob's wife, Marie-Paule, looks as if she has stepped out of a Hollywood film. When we part, the vivacious blonde kisses me directly on the mouth. Granted, she has the perfect lips for this.

Back in the Hyatt Regency I collapse in the bar with a journalist from *Le Figaro*. The French may be chauvinistic, but this chap is clearly pro-Gallo. He appears to have been a neurological psychiatrist who threw himself into scientific journalism, and he finds it an honor to be able to mingle for a week with the international protagonists of AIDS research, as at present. That much we have in common, I think silently.

On my last day in Washington I visit my old friend, biologist Brian Robbins, and his wife, Norrie. The couple live in a leafy outer suburb of the capital city. Norrie, an enthusiastic geologist and feminist, fetches me at the kiss-and-ride zone of the metro station. I receive the kiss promptly, and the ride is fulfilled in her beautiful new car. When I see her from time to time, Norrie asks me to tell stories. Coincidentally, Bob also asked me to do that yesterday, when we dallied in his garden next to the Italian delicacies. "Sorry I don't have a microphone, otherwise you could tell your stories to everyone here." I am a scientist, but perhaps I find it more stimulating to talk about science and interesting questions in general than to experience the "aha!" moment of discovery. Bob and Norrie are the same. I can listen, certainly, but I also find it wonderful when people hang on my lips. Hence I am driven to teach, to tinker about with cameras, and to hobnob willingly with people who have the gift of language. For the same pay it could have been the other way around: Laurie

Garrett might have finished her PhD, I could have been interviewing her in that Italian restaurant, and afterward I may have written *The Coming Plague*.

Late 1993

The old structure of the ITM, with fifteen small and relatively independent services, no longer appears to be suited to the conduct of current science, multidisciplinary collaboration, and cost-effective management. From now on there are five departments. My own, Microbiology, comprises four units that work together closely: AIDS/HIV (under the leadership of Marie Laga), mycobacteriology (Françoise Portaels), immunology (Luc Kestens), and virology, a unit that I lead myself, in combination with management of the department as a whole.

Our units are administratively, materially, and organizationally supported by our multitasking octopus Jan Vielfont. He keeps his eyes strictly focused on how much money we can spend on each project and spares the team mountains of paperwork. Thus the investigators can apply themselves to research and study. In the beginning Jan experiences contrary winds from the central ITM administration, which doesn't see why we need our own supporting administrator, but later all four of the other departments will follow suit and each will appoint its own Jan.

In the years that follow my division will grow to a group of about seventy people, some of whom are based overseas. I present mottos to my staff such as "Make your work an international network," because "You don't do science alone." I encourage them to maintain good contact with colleagues inside and outside the Institute for Tropical Medicine, and also to work for a time in other laboratories to pick up new ideas. I send my PhD students—the oxygen of my service, I call them—to Bob Gallo's annual AIDS conference or I let them give talks at other congresses and encourage them to network. That delivers. In this way I arrange a doctoral post for the Cameroonian student John Nkengasong at the CDC in Atlanta, after which the Americans offer him a career. John is now the adjunct director of African HIV research at the CDC. He still gives me a phone call around New Year to hear how things are going.

Good contacts are gold, in the academic world too. Over the years I have built an exceptionally fruitful collaboration with Francine Mc-Cutchan, an outstanding molecular biologist who first worked for the US

Military HIV/AIDS Research Program in Washington and who now works for the Bill and Melinda Gates Foundation. I put her in contact with my doctoral student Joost Louwagie, who supported Francine in her investigation of the genetic structure of HIV. During the same period my departmental colleagues Dr. Wouter Janssens and engineer Leo Heyndrickx record first class results in the area of molecular biology. With the PCR technique they analyze living viruses and put the genetic variability of many HIV clades (subtypes) on the map.

I had noted during my expeditions in the African jungle that team leadership suits me well. I make it a sport to allow the abilities of each individual to blossom, so that the whole team improves from this. Professor Stefaan Pattyn was my scientific mentor, but I never wanted to copy his top-down style of leadership—which was wholly normal in the 70s and 80s. Pattyn was moreover a genuine "Out of Africa" character who was accustomed to command, and who would give you an A-to-Z of what you must do. He even had an intercom with which he could give orders to Peter Piot and myself when we were busy in the lab. Pattyn earned respect with his encyclopedic knowledge and sharp understanding, but he was not easy to get along with.

I do try to be easygoing and tackle things resolutely but calmly. I try to be as diplomatic as possible and to show interest in what my colleagues are doing. The fact that I remain unstirred by all circumstances now and then drives Dina around the bend, but it is valued at work. I never lift my voice. At the same time I insist that protocols are observed—that is what they are there for—but simultaneously I listen to the improvements that people suggest. I expect my colleagues to work precisely and to show enthusiasm. And whosoever may be sick of his career, I would prefer to leave.

To keep a lid on things, I sketch cartoons. During dry-as-dust meetings I caricature someone's head, for example, so that everyone can have a laugh and the remainder of the meeting delivers that much more. My secretary, Ciska Maeckelbergh, says that she can reconstruct her life with the aid of my cartoons, from getting her driver's license, through hobbies such as ballet, to her separation. Sometimes she is depressed and I can pep her up with a cartoon, which brings me satisfaction, too.

I am also maximally accessible to my people. The door of my office always stands wide open. That has another advantage. Above my door I have installed a crossbar from which I can dangle to stretch my bad back whenever it gives me pain. "Our boss is hanging out again," my colleagues say. Funny, hey? The doubtful look in the eyes of passers-by when I greet them as I am swinging slowly to-and-fro in my doorway, amuses me greatly. But I transform with great ease from jester into professor. Many journalists have experienced me cracking jokes liberally before their interviews, and then, when the recording starts, I suddenly become dead serious.

I also go to lecture at the Vrije Universiteit Brussel (Free University of Brussels), where for nearly ten years I have animated young people with tales about infectious diseases and AIDS research. Already during my doctoral years I liked giving lessons and enjoyed the company of students, and at the VUB that still appears to be the case.

Saturday, December 25, 1993

In the postbox sits a manuscript: *Summer Idyll* by Bill Close. The Dutch translation is there next to it: *Zuster Veronica, het drama van Yambuku*. A fine Christmas present.

1994

The ITM and Innogenetics successfully map the entire genetic structure of HIV-1-ANT-70. The more we know about the genetic code of the virus, the greater the chance that someday we will make a vaccine against AIDS. Colleague Martine Peeters is on the trail of a third monkey AIDS virus. In Ivory Coast we find patients who are infected with both HIV-1 and HIV-2, and colleague Katrien Fransen applies herself to the further development of PCR technology to unlock the secrets of HIV.

I give an interview to *De Standaard* as an introduction to the appearance of *Zuster Veronica, het drama van Yambuku*. Readers are receiving the novel as an insert in the newspaper.

Together with a delegation from the Ministry of Foreign Trade, I go to a congress in Oxford. At the airport I am introduced to Paul the Baron Janssen of Janssen Pharmaceutica, Nobel Prize Winner Christian de Duve and his charming wife, and to several captains of industry of Fabrimetal and Gevaert. I also meet colleagues from Innogenetics. A bus arranged by the Belgian Embassy in London takes us to the center of the centuries-old university town. Oxford breathes serenity and peace. I lodge in the magnificently sited seventeenth century Bath Place Hotel.

The following day I hasten to breakfast at the prestigious The Randolph, where I am introduced to Belgium's Prince Filip, the permanent leader of the delegation for foreign trade since his father unexpectedly became king the previous year. One of the top officials in foreign trade is Mr. Roelants. It seems that he knows Ariadne Petridis, the niece of my wife. Ariadne works as a young diplomat in the Belgian Embassy in Kinshasa. A striking coincidence.

As we leave the hotel the prince walks beside me. I tell him about the visit that his aunt, Queen Fabiola, made to the ITM a year earlier. Boudewijn was still alive then. I invite Filip to come and visit us at the institute also. It would fascinate him, he says, and he tells me of his trip to West Africa where he saw with his own eyes the devastation wrought by AIDS. He visited two hospitals there and learned that almost half the mortalities were due to AIDS. While we wander farther a security guard asks him to return to the royal limo. "Tot ziens, monseigneur," I call after him and continue on my way at the side of Paul Janssen. During the short bus trip to the Radcliff Hospital, he questions me about recent developments in the group of O-HIV-1 viruses, but Dr. Paul is also interested in my Ebola studies. He looks up, surprised, when I tell him that we at the ITM were the first scientists to isolate Ebola, from the blood of a Belgian mission sister.

End of 1994–Beginning of 1995

In the Ivory Coast a 34-year-old Swiss behavioral biologist picks up an infection after investigating a dead chimpanzee that she encounters in the Taï National Park on November 16, 1994. During the autopsy she wears a pair of old household gloves. In her account no cut or prick incidents are reported. She probably becomes infected by minuscule droplets of chimpanzee blood that enter her eyes, mouth, or nose. Eight days after this she develops symptoms of what will later be diagnosed as Ebola. She admits herself to the main hospital in Abidjan and is repatriated from there by air ambulance to Switzerland, where she receives fluid and electrolyte replacement therapy. The woman is not quarantined, but no secondary infections follow. After two weeks she is discharged, although it will take another four weeks for her to recover completely.

This is the first and so far the only case of a new variant, Taï Forest Ebola virus. Of course, this type of Ebola virus has nothing to do with

Thailand, but is named after a nature reserve of 330 square kilometers, which as one of the last pieces of virgin tropical rainforest in West Africa shelters eleven higher primates, and also threatened animals such as the dwarf hippopotamus.

Hemorrhagic fever breaks out in Mékouka and other gold mining villages hidden deep in the rainforest of Northeast Gabon, and 52 people become infected—31 do not survive. That puts the case fatality rate at 60 percent. The clinical symptoms at first make one think of yellow fever, but after the event, Ebola is found in some of the tested sera, not only in patients but also this time in people who have not been ill. In February and July 1996 there are once again outbreaks of Ebola in the same area.

1995

We discover that the genetic diversity of HIV in Africa is much greater than we had hitherto thought. It is apparent that HIV adopts all sorts of forms. There are many viruses that continually change their structure, for example the flu virus. This makes it very difficult to make a vaccine against these viruses. And if it succeeds, then it still doesn't help against every exposure. The flu epidemic in Belgium in early 2015 was the largest in years, because it had been thought before the vaccine was produced that different strains would be present.

With HIV the situation is even more complex. Millions of patients have different strains of HIV in their blood. On top of this, the virus still mutates in each patient independently. At any instant a patient can thus be the carrier of a series of HIV strains. Add up all that and you get a gigantic swarm of different HIV viruses. In this way, an estimated 10^{14} to 10^{18} genomic combinations of HIV exist. The last figure is 1 with eighteen zeroes, or one trillion combinations, for those of us to whom such a number is meaningful. It comes down to the fact that one must check all these viruses to find out which parts they have in common. If you have the right primers (keys to find pieces of genetic material), then in principle you can use PCR to reveal all the possible strains of HIV.

When the Ebola virus once again explodes forcefully in Zaïre, I experience the frustration of a lifetime. On April 4, 1995, everything begins. On this day a laboratory technician in the hospital at Kikwit, a town of half a million residents 350 kilometers east of Kinshasa, develops fever and bloody diarrhea. A week later he undergoes an operation because the doctors think that he has a perforated intestine. On April 14 several of his hospital colleagues have the same symptoms. The disease then spreads to Mosango, 100 kilometers from Kikwit.

On May 6 the American Embassy in Kinshasa alerts the CDC in Atlanta and the WHO. Three days later the CDC receives 14 blood samples from Zaïre, although that did not go without incident. This I will explain shortly. Regardless, it rapidly appears that this is no limited outbreak with several tens of cases, but a full-blown epidemic with roughly the same number of infections as at Yambuku: 315 people pick up the virus and more than 80 percent of the patients will lose the battle.

The difference is that this time the epidemic is not only being covered by the Yambuku Broadcasting System and reporter van der Groen, but by almost all the big media concerns. In the middle of May, Ebola is genuine world news for the first time in its history. In 1976 the hullaballoo was chiefly limited to scientists and other people who were professionally involved in health care in Africa. A fair number of articles appeared in newspapers, but these did not linger in collective memory. Now, virtually every citizen receives TV images of what Ebola can do. The word "Ebola" becomes a definitive synonym for a terrifying viral disease that transforms your innards into a bloody pulp and kills you in a couple of days. This is the crude picture that the word "Ebola" conjures in many people.

People such as I, who have been following Ebola for almost two decades, think that in the meantime a few things would have changed. For example, that the transport of Ebola samples is now subject to strict conditions. Nothing could be further from the truth.

On Saturday May 6 at 06:00, Sabena flight 556 from Kinshasa lands at Zaventem. Among the passengers there is a woman with a coldbox in which there are fourteen vials of blood. No one checks her baggage and she also walks through passport control without questions. In the arrivals hall she is awaited by Dr. Johan Van Mullem of the Belgian Development Cooperation. The day before Dr. Van Mullem received a brief phone call from Kinshasa and learned that blood samples from Kikwit are on their way. Because the offices of the Belgian Development Cooperation are closed on the weekend, he drives with a coldbox of lethal viruses to his apartment in Leuven, where he takes the thing to his living room.

At 08:00 Johan phones his colleague Simon Van Nieuwenhove. At the Cooperation, Simon presides over the medical projects in Central Africa, but we also know him as an old stalwart of 1976. At that time Simon drove all on his own from Yambuku to the southern Sudan, to check if the outbreak there was related to the one in Yambuku. No, it wasn't—so it appeared later, but Simon didn't know that when he tackled the route of about 500 kilometers to the Sudan, and at each moment he could have come face to face with the deadly disease. Simon was also on the spot during the Ebola outbreak of 1979 in the Sudan, and for several days he has been keeping abreast of the news from Kikwit.

Van Nieuwenhove goes to pick up the coldbox from Van Mullem and sets course for Antwerp. He had been able to track down my private telephone number. We reached an agreement at the ITM. Dina can take a lot, but I fear that she will never forgive me for bringing Ebola home. Of course I know what is going on in Kikwit, but I am no longer working with Ebola on a daily basis. Obviously, in Kinshasa Dr. Jean-Jacques Muyembe thinks that I am.

During the middle of the day I take the coldbox from Simon. The absurdness of the situation suddenly strikes me. Nineteen years after we at the ITM confronted and handled something that appeared to be new and deadly, which would directly motivate the construction of our maximum security lab, I have the unique opportunity to be the first to detect Ebola in the samples from a new outbreak, but I must send everything at once to the CDC in Atlanta, according to the rules of the WHO—despite the fact that our current lab is a hundred times better equipped for this work than it was in 1976. The management of the ITM and the WHO would instantly have me institutionalized if they knew that in 1995 I am now going to do what I then did with the blue thermos flask. And on top of this we know that in the intervening period we actually had a maximum security lab.

I fear that in Kinshasa they clearly hadn't reckoned with the fact that this cargo would be underway for longer than planned, so I open the box. Just as well, because the dry ice that must keep the samples cold needs to be replaced. That could be a problem, as the ITM's supply is almost finished—in such a case the coincidence seems merciless. It is Saturday afternoon. Tomorrow is Sunday, and Monday is a national holiday. Grrr. And how will I be able to arrange a courier service to get the high-risk package to America as soon as possible?

While awaiting a creative solution I stow the specimens in the deep freezer. That buys me some time, but naturally I don't wish to be the cause of delay in this story and so I start phoning like a madman. From pillar to post and back again, I eventually locate dry ice and soon after I find Federal Express ready to transport the coldbox to the United States. However, I must immediately jump in my car and chase to the FedEx offices. I drive as fast as circumstances permit, in my turn following Johan Van Mullem and Simon Van Nieuwenhove in carrying Ebola blood through a Flemish city. Arriving at FedEx I hear from the friendly but conscientious counter clerk that I must first fill in a risk assessment form. That evening, despite all my scientific frustration, I can look back tired but satisfied on the successful completion of a "mission impossible," and I let the CDC in Atlanta know that they must expect fourteen specimens from what appears to be a new Ebola epidemic.

Following investigation, the outbreak of 1995 in Kikwit and its environs appears to have begun locally with an index patient who lived at the edge of town, next to a forest. On this occasion Bob Colebunders of the ITM travels to the battlefield to conduct research. To each his Ebola time. Shortly after the outbreak the WHO establishes a special Emergency Division under the leadership of my good friend Dave Heymann. From now on the unit will be permanently ready to intervene in an outbreak of hemorrhagic fever, whatever virus is responsible. The world is constantly improving, but now and then it requires a disaster for important steps to be taken.

First Half of 1996

Near Mayibout in Gabon, hunters find a dead chimpanzee in a forest. The dead animal is cut open. Nineteen people who are involved become sick and pass on the infection to relatives. The cause is Ebola, with 37 cases and 21 deaths this time, or a CFR of 57 percent.

The advent of combination therapy against HIV and AIDS is a breakthrough. In HAART (Highly Active AntiRetroviral Therapy) different drugs are combined. This will dramatically improve the prognosis for infected people, so much so that at the beginning of the 21st century, HIV infection will become a chronic ailment with which you can grow old, provided you have access to the right medication. HIV-inhibiting agents were also developed before 1996, but these were taken separately and did not work for long. They also often had serious side effects. HAART does not have these disadvantages. Mostly three different medicines are combined that counteract the multiplication of the virus in as many different ways.

Wednesday, September 4–Saturday, September 7, 1996

Twenty years after we discovered a new, Marburg-like virus at the Institute for Tropical Medicine, I can invite the cream of the international Ebola researchers to an Ebola colloquium in Antwerp. Science journalist Marc Geenen, who I have known for years and who has become a friend, makes a documentary using the footage that I shot for the erstwhile YBS. Marc also supplies an English language commentary, so that the congress attendees gain a comprehensive overview of what played out in Yambuku.

Following the congress there is an afterparty at our house in Kontich. Ciska, my secretary, is present, along with Joel Breman, the divorced couple Patricia Webb and Karl Johnson with their new partners, and a whole lot of other old and new warriors on the Ebola front. It seems odd to see them all together in our house, but it feels good—for them too, apparently, because that night a romance blossoms between two researchers. Thus, besides being a mistress and a marriage breaker, now and then Ebola can also be a matchmaker.

1996–1997

Ebola breaks out in the region of Booué in Gabon, and after patients have been transported it also strikes people in the capital city Libreville. The index case is a hunter who lived in a forest camp, where a dead, infected chimpanzee is found. Sixty-nine people are infected, of whom 45 die (a CFR of 75 percent).

A health worker who has handled Ebola patients travels to Johannesburg in South Africa. The man himself survives, but the nurse who looks after him in Johannesburg is infected and dies.

Sunday, May 18, 1997

Under the leadership of Laurent-Désiré Kabila, the Alliance of Democratic Forces captures the Zairean capital Kinshasa. The combatants are jubilantly received by the population. The "happy entrance" is the culmination of a campaign that began half a year earlier. A few days later Kabila names himself president. The reign of dictator Mobutu Sese Seko is at last over and Zaïre will again be known as Congo, but peace and stability do not come to that land. The rebuilding of the health infrastructure also remains undone. Malaria, diarrhea, and respiratory infections—diseases that are perfectly treatable—flourish abundantly. Particularly in the east of the country, where a bloody civil war has raged, people can scarcely find proper health care. Childhood mortality is six times higher than normal, even for a developing country.

Elsewhere in the Congo it is not much better. Most Congolese are so poor that they only seek help when it is already too late. Donor country Belgium helps but doesn't believe in a policy of free assistance; she stresses that patients should contribute something toward their own medical care and urges that medical aid funds be created. According to aid organizations such as Doctors Without Borders (DWB), due to this the poorest people who have the greatest need remain bereft of all medical help.

Regardless of this, health care is hopelessly underfunded in the Congo. Annually the government spends about one euro per head on health care while international donors spend two to three euros, but according to the WHO at least 16 euros are required per year per resident. Moreover, doctors and other health workers are appallingly paid. According to DWB, they earn five to twenty euros per month in the rural areas. What can one do? Before independence the Congo had just about the best health service in the whole of Africa, but not a single aspect of the colonialism upon which that system was based was maintained. Under Mobutu, health care imploded entirely. Aid organizations and clerics such as the Flemish sisters of Yambuku did what they could, but due to a lack of planning and support on the part of the Zairean government—and despite all their fine intentions—they sometimes kicked the ball far wide of the goal. This we had seen in 1976.

Now there is the exchange of power from Mobutu to Kabila, but I suffer no illusions. I am a born optimist and I try to see the good in everything and everyone—if someone later remembers anything about me when I am no longer here, then hopefully that will be it—but one becomes despondent about what has happened in the Congo, then Zaïre, and thereafter Congo, and indeed in almost the whole of sub-Saharan Africa.

Congo is bursting with mineral wealth; above ground it has more landscape and natural beauty than the most demanding tourist could possibly wish for, and it is populated by people who seem to be born with laughter and an inextinguishable lust for life in their bones. But we are not succeeding in supporting them and at the same time treating them with respect, and they are not succeeding in organizing themselves democratically and in rooting out institutionalized corruption and continuing violence. Many African countries lie sick in the same bed. The continent resembles a single great termite mound that devours urgent aid, and I creep with a magnifying glass into all its passages and corners, searching for viruses.

August 1999

At Bob Gallo's annual congress, one of my master's degree students, Fatim Cham, gives a first-class presentation. Bob congratulates her publicly. I glow with atavistic pride. Her brother Hassan, a computer specialist, would like to see his sister at work and traveled with her to the US. He is just as much impressed with Fatim's presentation as I am. He normally looks after his sister's children, which gives her the opportunity to devote much of her time to science—yet another proof that as a researcher you are only worth as much as the support that you receive from others.

Sunday, July 9– Friday, July 14, 2000

Sixteen years after the first AIDS conference in South Africa, it is still very difficult for President Thabo Mbeki to accept the "western" scientific knowledge of AIDS. I am there at an international AIDS conference in the coastal city of Durban when, during a speech, Mbeki's Minister of Health Manto Tshabalala-Msimang announces before an audience of 10,000 people that "South Africa will find its knowledge about antiviral products to fight AIDS in nature. We have all the herbs and plants that are required and with them we will complete the task."

In the end his predecessor Nelson Mandela, who retired from the presidency a year earlier, tries to rectify the murderous nonsense of Mbeki a little. With great diplomacy he thanks foreign countries for their dedication to the battle against the AIDS catastrophe that has hit his country. "In close cooperation with the international community, South Africa will fight to see that those infected with HIV will receive humane treatment," promises the legendary anti-apartheid activist. A later study by the Harvard Medical School will estimate that Mbeki's categorical refusal to permit government support for the right HIV medication allowed 330,000 South Africans to die from AIDS, unnecessarily.

Monday, September 4, 2000

When I set to work more than twenty years earlier in my maximum security lab, we worked according to self-imposed quality standards. In those days we were just a small team. Now our department counts too many people to allow each person to be closely followed and, if needs be, managed. It has therefore become essential to follow detailed protocols that have been written so as to leave no room for interpretation and that can be carried out by suitably qualified staff.

Katrien Fransen stands guard over our quality policy—with results, because the ITM's AIDS reference lab receives the official accreditation certificate awarded by the quality committee of the World Health Organization. In its report the committee states that it "is impressed by the quality, the knowledge and dedication of each of the staff members of the AIDS reference lab. It has been clearly demonstrated that the personnel work excellently and in an accurate manner they do what is prescribed in the standard work procedures." So we hear it from an independent source. It is a feather in the cap for all my staff and especially for Katrien Fransen. As an AIDS reference lab we are also part of the UNAIDS HIV network that my former colleague Peter Piot has been leading for a while.

2000–2001

Ebola breaks out in Uganda. Transmission of the virus occurs especially during funeral rites and medical care of the patients. 425 people are infected, of whom 224 lose the battle (a fatality rate of 53 percent).

October 2001–July 2002

Ebola strikes in the border area of Gabon (53 deaths) and Congo-Brazzaville (43 deaths). This is the first outbreak in Congo-Brazzaville. The CFR is 79 percent.

2001

In the ITM we have an enormous variety of HIV types, thanks to the sera of infected people who come through our doors daily. In the sera there are antibodies. The question is whether the antibodies from one person can neutralize the HIV viruses of another. If we could find a serum with which we could knock out twenty or thirty HIV types that differ significantly genetically, then we would further investigate the infected person, because this means that his antibodies contain some part that fits many virus types.

I try to explain this "broad spectrum neutralization" to my students graphically. Assume, I say, that you are all HIV viruses that differ from each other, and I want to make a vaccine that can wipe out each of you. That isn't easy, because I must find an element that you all have in common. Now look around carefully, and feel your nose. Is there someone whose nose has a square tip? No one! You have nose tips that all differ from each other a little. But perhaps among all the different nose tips there is a little point, a truly small part, that is nonetheless the same in every single one of you, thus in each virus. We now possess techniques that mean that we are no longer obliged to make a whole nose tip, but can make just a little part of it. That little part we use in a vaccine. In this way, we may be able to produce antibodies that attack a common point among all the millions of HIV strains.

Day and night we do neutralization tests at the ITM. We develop mathematical models to analyze all the data because if we have thirty different sera from thirty different patients with thirty different viruses, this yields an enormous table. Phillipe Nyambi, a very talented doctoral student from Cameroon, undertook fascinating doctoral research on this subject, in close cooperation with Paul Lewi, a top researcher at Janssen Pharmaceutica. I also gave Phillipe the chance to continue this type of work with my colleague Professor Susan Zolla-Pazner in New York, where he further developed into an excellent HIV researcher.

My department is subjected to an external audit for the period 1996–2000. I must appear before a jury on which top brass such as the boss of Janssen Pharmaceutica sit. How I admire this man. If you bump into him, he promptly asks you what news you have. One must then dish up something that Dr. Paul hasn't yet read or heard, and in this regard it is easy to disappoint him. When the jury's report is in front of my eyes, I convene my staff. The Microbiology Department of the ITM is among the top 20 worldwide! Not only our laboratory but also our field work is of international standard, finds the external panel. Thanks to the outstanding report we also attract more money.

2002

I turn 60. My official time at the ITM is over. I can still work somewhat longer—and indeed I will—but to keep on my toes chasing everything and to surf all the congresses—this must be scaled back. And the hunt for cents and the daily drive to lift the Microbiology service to a higher level and keep it there, this will shortly be the work of others. In the meantime we once again honor our role as a "WHO customer service." We publish a comparative study of eight diagnostic tests of the ELISA type and fourteen rapid tests. To do this we gathered and analyzed a good two hundred sera from all over the world, most of them from Africa.

December 2002–April 2003

A new Ebola outbreak in Congo-Brazzaville. Of the 143 infected people, 128 die. CFR: 90 percent.

Wednesday, January 1, 2003

Dina and I become pensioners. For forty years she has as good as stood alone. Shopping, cooking, caring for our daughters, washing, and the household chores. And if I was ever at home, she had to hear my stories about viruses. Because if I read a book or an article that inspires me, then I must talk about it with others. Frequently Dina was the victim. She is stronger than I, and indeed will remain so… I realize that she has sacrificed herself for our family. She is clever, perhaps cleverer than I, and has a brilliant academic record—she could undoubtedly have taken it further. That I am now writing these words is not adequate recompense. She had to put up with me, much more than I with her. It was often a case of dancing on a tightrope but—and this we did well together—the tolerance and will of both of us to find a consensus maintained the "unity in diversity."

Meanwhile I have become a grandfather with two granddaughters, and my pension gives me something of a margin to assume the role of "opa." But I remain busy as a professor, adviser, and guest speaker at congresses. Happily so. There is still too much drive in me to turn the setting of my life back to the pilot flame. I will certainly continue to follow Ebola and HIV. In this regard the Internet could not have arrived at a better time. I now pursue my hobbies in cyberspace too, and note the essentials of all the scientific goodies in my blogs.

November–December 2003

Seven months after the previous outbreak, Ebola again appears in Congo-Brazzaville, with 35 victims, of whom 29 die. CFR: 83 percent.

2003

We check to see if we can find antibodies that can neutralize a great variety of HIV types in the sera of people who have been experimentally treated with an HIV vaccine.

2004

Ebola strikes South Sudan. It later transpires that certain suspected Ebola cases were actually victims of measles. The WHO counts 17 cases and seven deaths.

The Virology service of the ITM is still searching for an HIV vaccine, in collaboration with Tibotec, a Flemish drug company that focuses on research and treatment of infectious diseases, especially HIV and hepatitis C.

2005–2008

I lecture on the subject "Molecular targets in infected cells" at the Vrije Universiteit Brussel (Free University of Brussels).

September–November 20, 2007

Ebola pops up in the western Congo. On November 20 the epidemic is extinguished. There are 264 cases and 187 deaths, or a CFR of 71 percent.

December 2007–January 2008

The first outbreak of BDBV* occurs in Bundibugyo, West Uganda, with 149 cases and 37 deaths, or a CFR of 25 percent. The Bundibugyo virus (pronunciation: boondy-bood-jeeyo) is an Ebola variant that, just like the "ordinary" Ebola virus, causes hemorrhagic fever. Including this newcomer there are thus five species of the genus of Ebola viruses: the Ebola virus itself (EBOV,* earlier named Zaïre Ebola or Ebola-Zaïre, the species that ravaged Yambuku), the Reston virus (RESTV,* the only species of Ebola that doesn't make people ill), the Sudan Ebola virus (SUDV*), the Taï Forest Ebola virus (TAFV*), and Bundibugyo virus (BDBV). I summarize them here because there is confusion about the name "Ebola" and the diverse species of this viral genus, which have had other names. In any case, these are the names and abbreviations as they are used by science in 2015. Five years after the outbreak in Uganda, Bundibugyo would also flare up in Congo, perhaps after people had butchered bushmeat and eaten it.

Monday, October 6, 2008

Luc Montagnier and Françoise Barré-Sinoussi are awarded the Nobel Prize for Physiology or Medicine for their discovery of the HIV virus. I would have found it fairer if Bob Gallo had been added to this duo, if only because it was due to Bob and his team that we had an HIV test relatively early.

December 2008–February 2009

Yet again an outbreak of Ebola in West Congo, in the same region as in late 2007. This time there are fewer cases, 32 to be precise, and "only" 14 deaths, or a CFR of 45 percent.

April 2010

In a good mood I take the train to Zaventem, where I simply and rapidly check in with my electronic ticket. I was pleasantly surprised when I received the invitation, together with another Ebola veteran—my friend Karl Johnson—to speak at a congress in Japan. Despite the second eruption yesterday of the Icelandic volcano Eyjafjallajökull (pronunciation: eeyafyat-la-yirkuk, although the last "k" sound is closer to the sound that you hear when you push a wall of your mouth to one side with your tongue and force air past it, for example to urge on your horse—or so this knowledge-hungry professor discovered online), my flight from Copenhagen to Tokyo is delayed for just two hours. I was dreadfully lucky because shortly after our departure Copenhagen Airport is closed to all traffic.

Twelve hours later I am in Japan, with a whopper of a jetlag. Of course, people are not made for such a complete attack on the wake-sleep cycle and interference with the biological clock. In evolutionary terms we hardly differ from our ancestors five thousand years ago. When they relocated their bodies had all the time required to adapt to the changing landscape, and it scarcely had any effect on their day-night rhythm. They saw day by day how the landscape in which they awoke followed the slow rhythm of the seasons. They traveled no faster than their limbs, their strongest sled dogs, or their fastest horses could carry them. In modern times, our day—especially if we travel east–west—can, with a single flight, be doubled or halved. After this we step into a new country or onto a new continent, where our senses are inundated with new impressions to process and a temperature difference that is sometimes 68 degrees.

But also for those who stay at home, a day in the 21st century is in no way comparable with a day in the Stone Age. We no longer wake with the first glimmer of dawn or with the birds that begin to twitter, but with an alarm radio that buzzes, or spews out music or talk about events. At breakfast we check our Twitter and Facebook services, our texts and our e-mails. On the way to work the car radio is on and we keep one eye on the announcements on our GPS and the other on the traffic and the hundreds of billboards, bulletin boards, and traffic signs that we pass. Everything screams for our attention, everywhere and at all times. Subsequently we spend the remainder of the day and the evening next to screens: computers, TVs, laptops, tablets, e-readers, smartphones. Older people like myself still keep these things separated, but the digital natives of the younger generations keep up with all the screens at the same time. They multitask, chat, and swipe their way through life, hustle the real world and cyberspace together without difficulty, and find it the most natural thing in the world.

All of it is splendid, because communications technology has made the whole planet accessible and our lives incredibly more interesting. That having been said, it would not surprise me if the epidemic levels that stress, burn-out, and psychological problems have attained cannot at least be partly ascribed to the uninterrupted bombardment of the human brain, which has remained stuck in prehistoric times morphologically. If you were to scan the brain of a Cro-Magnon baby and compare it with that of a tiny tot today, our neurologists would probably be unable to distinguish between them. By contrast, the worlds of these two infants differ as day differs from night. Perhaps this is why we sometimes long for water, air, earth, and fire. The primal elements set us at rest. Listening to waves washing onto a beach or the tapping of the rain on a window, watching clouds slowly drift past or a clear starry heaven, a thunderstorm, lightning, to pull one's hands through soil, to garden, to fashion with clay, to stare into the hearth or an open fire, just to dunk one's head beneath the water in a bathtub, swimming pool, or the sea. To shelter from a downpour during a mountain hike and a little later, as the sun again breaks through, to smell the steaming rainwater on the rocks... all that fascinates, mesmerizes, hypnotizes. The primal elements remind us of the décor that surrounded our ancestors for hundreds of thousands of years. They give us a break from the bombardment of the senses and give our overstuffed brains the opportunity to switch back off. Also in their most sublime form—art—they have the ability to restore a person to his natural self.

I travel from Osaka to Tokyo. In the garden of the museum where I am taking a breather, a woman is continuously walking around with a brush and pan. Not a single leaf lies on the ground. The near-clinical tidiness of Japan surprises me yet again. I buy a ticket for the limousine bus that will take me to the Sheraton Myaki Hotel, and I am asked extremely politely to proceed to bus platform 13. A little later a friendly man comes to meet me. He bows respectfully and checks my ticket, to see if I am indeed on the right platform. He numbers my suitcase and gives me a copy of that number. The bus is expected at 12:30. It is 29 minutes and 55 seconds after twelve when it arrives. While our cases are loaded, the passengers climb aboard. As the bus departs the luggage boys stand trimly in a row to bow farewell synchronously.

We pass a sparkling fountain in Ueno Park that I visited earlier. I had been impressed when I learned the tale of Tatsuo Yamamoto there. On August 6, 1945, the day that the atom bomb fell on Hiroshima, Tatsuo sat in a train going to the barracks where he was lodged. Because his uncle had a bookshop in the center of Hiroshima, Tatsuo interrupted his journey to search for his relative. Just like the rest of the city, the bookshop appeared to be burnt to the ground. The ruins of the house still smoldered. Tatsuo found a little hand warmer and with it he managed to carry a flame from the rubble back home. There, he kept the fire burning in a family shrine for 23 years, until the story became known in 1968 and with Tatsuo's flame other memorial fires were lit, in diverse monuments. Since that time a fire has been burning uninterruptedly in Ueno Park.

I was barely three years old, I reflect, when the bomb dropped on Hiroshima and three days later another fell on Nagasaki. At least 129,000 people died in the bombing, and afterwards just as many through radiation sickness and cancer. In other words a CFR of 200 percent.

That evening at the hotel I create an e-mail address. My password is "ebolaa." Dina, on the other side of the globe, answers immediately.

At the start of the Fifth International Symposium on Filoviruses I encounter fewer old acquaintances and more new faces, who for their part do not know who I am. Happily, Karl Johnson is still there. In the program it states that Karl and I are giving a presentation with the title "How two virus fighters became Ebola freaks," but Karl is lying in bed sweating out the flu. So I am on my own. My partly improvised talk falls on eager ears. The circumstances in which we handled a lethally dangerous virus, and especially the tale about the thermos flask, sound prehistoric to the young virological new guard.

During the later dinner I sit between young congress delegates. When I ask who was born before 1976, no one raises a hand. I was 34 when I saw the Ebola virus under the electron microscope for the first time in that year, and my colleague Karl Johnson was 47. At 81 years old he is now the oldest congress participant, and I am next in line.

The next day it is again so bitterly cold that I do not have the oomph to travel to Katsu, my favorite sushi restaurant in Tokyo. So I will make do with the sushi bar of the hotel. It costs me 7,500 yen, which is roughly 60 euros. That is three times more expensive than Katsu. Before going to sleep I check out in advance and book my limo bus to Narita Airport. My nose is dripping and my sinuses are blocked. A little gift from Karl.

Tuesday, December 4, 2012
An old-age home now stands on the site of the maternity clinic where I was born. There my beloved mother dies at the age of 98. For her the circle has closed.

June–December 2012
June through August, there is an outbreak of Sudan Ebola in the district of Kibaale, Uganda. In July there are 14 deaths, 12 of the victims belonging to the same family. The government tries to trace people who are possibly infected and advises the population to avoid bodily contact and not to bury dead family members themselves. By August 12 the disease has been contained. It was the third outbreak in Uganda in twelve years. That of 2000, with 425 infections and 224 deaths, was the most deadly. There is an outbreak of Bungibugyo in the eastern Congo starting in June and

ending in November. The WHO fears that the virus will find its way to the big Congolese towns, which ultimately does not happen. There are 77 cases and 36 deaths, or a CFR of 47 percent.

The classic Ebola picture has gradually become clear. If we neglect the extremely rare Taï Forest Ebola and the fairly innocuous Reston virus, we have had seventeen Ebola outbreaks since 1976, thus a little less than one every two years, all of them fairly central in Africa (in Sudan, Gabon, Uganda, Congo-Brazzaville, and Congo/Zaïre), repeatedly with several tens to several hundred victims and a case fatality ratio of 25 to 90 percent, mostly about 70 percent. Roughly guessing, the next outbreak will occur in 2014 in the Congo, will infect 250 people and kill approximately 175 of them. We shall see.

CHAPTER 6

2013 – 2015

PUSHED TO THE LIMIT AND BEYOND

How an old virus outwits a new region
with old traditions and leaves behind
its greatest footprint ever.

Friday, December 6, 2013

Saint Nicholas is probably too busy with the children of Flanders and the Netherlands, otherwise he would have done something about the death of little Emile, a toddler of two in the Guinean town of Meliandou. And it doesn't stop there, because shortly after this his mother, older sister, and grandmother die too. Outside the family and the immediate environment no one bats an eyelid at these deaths. Here, people are continually dying from anything and everything. Malaria, cholera, HIV... you name it. In Guinea healthcare is rare and expensive. People have poor access to potable water. The income generated by the exploitation of bauxite reaches everyone except the Guineans themselves. The only thing that is properly shared in this country is poverty.

Just one year later, in October 2014, researchers will identify Emile Ouamouno as patient zero* in what had by that time become the greatest Ebola epidemic in history. This could be ascribed to the fact that a disease that had never before been identified in West Africa spread among humble households in the south of Guinea and could circulate for three months virtually unnoticed, overshadowed by all the health problems of this wretchedly poor country on the Atlantic Ocean, crippled by criminality and corruption.

Tuesday, March 18, 2014

The Guinean Ministry of Public Health announces the outbreak of a hemorrhagic fever that "strikes like lightning." The government could nevertheless have already been hearing the thunder for quite a while, because to date there have been 35 cases, with 23 patients now dead. Several days later Doctors Without Borders is on the spot. The organization erects isolation infrastructure in the Guéckédou prefecture and establishes a second aid center in Macenta.

Monday, March 24, 2014

It appears that in Liberia, too, near the border with Guinea, five people have died from hemorrhagic fever. Children's rights organization UNICEF fears an epidemic.

In Guinea the disease has meanwhile reached the capital city Conakry. The World Health Organization sends experts.

Tuesday, March 25, 2014

The American Centers for Disease Control and Prevention (CDC) issues a first report: 86 cases in Guinea, among which there are 59 deaths, or a case fatality rate of 68.5 percent. The cause is probably the Zaïre Ebola virus (EBOV).

The WHO also publishes its first official report on the epidemic in West Africa: the laboratories of the Institut Pasteur in Dakar (Senegal) and the Centre International de Recherche en Infectiologie in Lyon, France, have found 13 cases of Ebola in total. The blood samples came from four districts of Guinea.

You might think that the figures of the CDC are in disagreement, but that is only to do with the difference between the number of cases determined in the field on the basis of outwardly visible clinical symptoms and the number of cases confirmed in the laboratory. This is normal in an epidemic. The lab confirmations trail behind the on-site estimates. In 1976, during the epidemic in and around Yambuku, this was also the case. And mistakes also happen.

In the beginning, when the cause isn't yet known, people are buried before blood is drawn from them. Other diseases are taken to be Ebola, while genuine cases are missed because people have malaria at the same time and a test is only available for this latter disease. Fatal cases are added to infections while they have already been counted as a part of these, and patients panic, flee, and die in solitude without ever becoming a statistic. There is overestimation as well as underestimation.

Only when the worst panic has subsided and the aid organizations are more or less standing on their feet, can the blood sampling and test infrastructure function at full speed and definite diagnoses be determined on a large scale. However, that happens not only at many different centers in the affected region, but also in other diverse countries. All the details must be compiled, such that the final report can often only be made months after the epidemic. And the final report, too, is never one hundred percent correct: data are lost, vials of blood are mislabeled, people are tested too early, some patients are tested several times. In short the case fatality rate, the ratio between the number of infected people and the number of deaths, must always be taken with a pinch of salt during an epidemic that is still raging.

Friday, March 28, 2014

In Conakry four infected people are placed in quarantine. The news sows panic in this city of a million souls. In south-eastern Guinea there are already 66 deaths.

Senegal closes its border with Guinea.

Ebola reports emanate from Sierra Leone, and in Liberia, too, there are two confirmed cases, in the regions of Lofa and Nimba. It is officially stated that the patient in Lofa has died.

Monday, March 31, 2014

New figures from the WHO: 70 deaths and 112 people who are possibly infected, almost all of them in Guinea. Doctors Without Borders expresses its alarm and concern about an "unprecedented spread."

In Yambuku the damage was limited to patients within a radius of a few tens of kilometers around the mission. Now there are Ebola cases in three countries simultaneously. If you look at Google Earth and zoom in on Meliandou, where the index patient Emile died, you will see that the village lies right next to the point where Guinea, Sierra Leone, and Liberia meet—these being three coastal states. The square of Sierra Leone, in the middle, is the smallest land of the three. In the east it borders on the significantly larger rectangle of Liberia. West of Sierra Leone lies the coast of Guinea, by far the largest country of this trio. In the hinterland to the east, the crescent of Guinea curls over Sierra Leone and Liberia like a wave washing over these two neighboring states.

The CDC sends a team of five from America to Guinea to support the Ministry of Public Health and the WHO.

Tuesday, April 1, 2014

Saudi Arabia lets it be known that no visa will be granted to Muslims from Sierra Leone, Liberia, or Guinea who wish to take part in the Haj at the beginning of October. Usually more than seven thousand West African pilgrims would come to Mecca.

Friday, April 4, 2014

Foreign mining companies in Guinea stop operations. Expats leave. In Conakry not even a third of the rooms in the luxurious business hotel Palm Camayenne are occupied.

A treatment center in the south-east of the country is attacked. Health workers in the West African region are encountering increasing hostility. People are fearful and suspicious.

In the whole region there are now more than 90 deaths, of which 86 are in Guinea.

In France the hospitals are in a state of preparedness in case a traveler from one of its former colonies imports the virus.

Tuesday, April 8, 2014

In Flanders, too, people gradually react. Two Flemish high school exchange students who are learning in Gambia return because "the whole of West Africa is a risk area."

Gambia? That is a fair distance from the "hot zone," and this little country is wholly surrounded by Senegal, where at present no cases have been announced.

Saturday, April 12, 2014

Two new cases in Liberia.

Monday, April 14, 2014

According to the Guinean Minister of Foreign Affairs the epidemic in his country is under control. The virus has claimed 101 lives from 157 people infected. All too soon it will be seen that the minister has cried victory too early.

Wednesday, April 30, 2014

The outbreak in Guinea is clearly not under control. A second wave is underway, with 221 confirmed cases—among which there are 26 health workers—and 146 deaths.

Monday, May 12, 2014

There are twelve new cases in the Guinean capital city Conakry.

Monday, May 26, 2014

The WHO announces the first Ebola deaths in Sierra Leone, in the district of Kailahun. Shortly thereafter this land closes its borders with Liberia and Guinea, and the first schools are shut.

Tuesday, June 17, 2014

Liberia announces that the virus has reached the capital city, Monrovia.

Friday, June 20, 2014

The WHO announces 158 cases in Sierra Leone.

Saturday, June 21, 2014

Doctors Without Borders says that the outbreak is spreading out of control and asks for help and intervention on a massive scale. There are already 350 deaths. If that is right, we are dealing with the most deadly Ebola outbreak ever and the sad record of Yambuku 1976 has been broken. Then, we recorded a final figure of 280 Ebola deaths.

Thursday, July 3, 2014

International health organizations and the governments of eleven African countries gather in the Ghanaian capital of Accra. They wish to monitor the situation better and collaborate on border control. One of the biggest problems is that the population in Guinea, Liberia, and Sierra Leone does not know this disease, says Doctors Without Borders (from Belgium).

Indeed, West Africa is not Congo, where six outbreaks have already occurred. We are inclined to bundle together all the countries south of the Sahara. This is because most of the people there are poor and black, and it is continuously warm—although in the mountains and in the whole of Southern Africa during the winter months from June to August, that can be wholly false—but furthermore Guinea differs just as much from Congo as Switzerland differs from Spain.

Friday, July 4, 2014

The epidemic in West Africa could "still last for months," says the WHO. The tally stands at 470 deaths. There are no concrete plans for radical regulations and there is no extra money.

I appear on the TV program *Terzake*. In this interview and the many that follow I stress that Ebola is also a "virus of fear" that gives rise to wrong decisions (stigmatization of people, the suspension of airlinks, etc.) and postpones the implementation of correct measures (lack of investment in basic health care and good public enlightenment). Anxiety spreads still quicker than the virus itself and is now present worldwide. It irritates me without end when I read that cured patients are shunned by their fellow villagers due to anxiety and ignorance.

Epidemics are statistics and statistics are people. We get to know more details about some of the chief players in this human drama. In the Liberian capital city of Monrovia there is a Patrick Sawyer, who lets his line manager at ArcelorMittal know that he may have been infected with Ebola. Patrick Oliver Sawyer is a 40-year-old American-Liberian advocate who lives in Minnesota but who is staying in Monrovia. He works for the Liberian Finance Ministry and for the steel sector. Patrick has had contact with his sister, who may have Ebola. ArcelorMittal takes the message seriously and sends Patrick to the Ministry of Public Health. There he is requested to stay at home and not to go to work for 28 days. This story will develop further.

Friday, July 11, 2014

The WHO has set up a monitoring and coordination center. According to the organization there are now 900 cases in West Africa. Doctors Without Borders speaks of a "race against time."

Tuesday, July 15, 2014

During the past week 68 new fatalities have been counted. That brings the total to 603.

Thursday, July 17, 2014

In Sierra Leone there are 442 infections. This land is catching up on Liberia and Guinea.

Sunday, July 20, 2014

In Liberia, advocate Patrick Sawyer has received permission from his other manager, the Finance Ministry, to attend a conference in the Nigerian town of Calabar. Sawyer thus disregards the advice of the Public Health Ministry to remain at home for a month. Although according to later declarations of eyewitnesses he is already visibly sick on this Sunday, at Monrovia's airport he obtains an ASKY Airlines flight to the Nigerian city of Lagos with a stopover in Togo, without a problem. When Sawyer lands in Lagos, he stumbles into the airport. One of the people who rushes to help him is 36-year-old Jatto Asihu Abdulqudir, an official of the Economic Community of West African States, who has been waiting for Sawyer. Abdulqudir takes Sawyer to a hospital. And with that, Ebola has invaded Nigeria. Sawyer is "patient zero" in this enormous land of 175 million inhabitants.

Thursday, July 24, 2014
Sawyer dies. Abdulqudir is quarantined.

Friday, July 25, 2014
The first case of Ebola occurs in Freetown, the capital city of Sierra Leone, where more than one million people live. A woman was removed from a hospital by her relatives. After the authorities distribute a "wanted" notice the woman wishes to return to the hospital, but she never reappears. The authorities request the population to report every suspected case with urgency.

Sunday, July 27, 2014
Liberia closes most of its border crossings. At the most important border posts aid centers are set up to detect cases and isolate them. Football matches are canceled.

Tuesday, July 29, 2014
In Sierra Leone Dr. Sheik Humarr Khan dies, a local hero in the fight against Ebola. In the Kenema Government Hospital he treated more than 100 Ebola patients.

The Nigerian air transport company Arik Air stops flying to Liberia.

Wednesday, July 30, 2014
Liberia, which has counted 129 deaths, closes all its universities and schools. In the border region with Sierra Leone and Guinea no more markets may be organized. Officials who do not have any task that is absolutely essential must stay at home for a month. The government requests aid from the international community.

Sierra Leone sends in troops to safeguard quarantines.

The European Commission, which had earlier donated two million euros for the fight against Ebola, gives another two million. The money is shared between the WHO, DWB, and the International Red Cross. To date, 1,201 cases of Ebola have been confirmed, of which 672 have died. According to Bart Janssens of the Belgian DWB, the figures are systematically underestimated.

The British Minister of Public Health Philip Hammond states that Ebola is a threat to his country.

The United States withdraws all Peace Corps volunteers from Ebola-stricken countries.

Thursday, July 31, 2014

The United States and France issue a travel warning for Guinea, Sierra Leone, and Liberia. The last time that the US issued a similar warning was during the outbreak of the SARS virus in Asia in 2003.

Ernest Bai Koroma, the president of Sierra Leone, declares a state of emergency and sends in troops to isolate disease hotspots in the east of the country.

My friend and former UNAIDS director Peter Piot advises that experimental vaccines and treatments be tried for "compassionate use, but also to discover if they work." Compassionate use* is an exceptional category of medical care under which treatment that isn't yet legal may nonetheless be used. I agree with Peter wholeheartedly. It is now or never if we wish to give experimental vaccines and antiviral agents a push.

For years several of my colleagues—albeit on a small scale—have been developing antiviral products and therapeutic vaccines that have delivered promising results in test animals. But to their great frustration they did not find a single business ready to explore these products further and test them on people. It easily takes a decade to make a usable vaccine. The investment is enormous, and due to the limited number of people infected with Ebola during those years, the pharmaceutical industry couldn't see how the financial returns could even cover the costs, let alone generate profit. But since this outbreak the situation has changed. There are more than enough infections, and the dollars no longer need to come from the pharmaceutical sector. International money donors also put funds at their disposal. The ideal Ebola vaccine* is preventive and therapeutic: it prevents the virus from infecting people but also stops anyone who has just been infected by the virus from developing the disease.

The WHO says that 75 million euros are needed to fight this outbreak.

Friday, August 1, 2014

There is consternation in the US social media now that it is known that two American aid workers have been infected by Ebola in West Africa and are going to be repatriated. The two people concerned are Kent Brantly, a 33-year-old mission doctor from the organization Samaritan's Purse, who was infected in Liberia, and Nancy Writebol, a nurse.

The American entertainment and hospitality magnate Donald Trump posts tweets such as "People that [sic] go to far away [sic] places to help out are great—but must suffer the consequences!" Many Americans react with indignation. Actress Whoopi Goldberg calls the tweets of her filthy rich compatriot idiotic.

In the whole region 1,323 people are now infected and 729 patients have lost their lives. The epidemic is getting out of hand but can still be stopped, says WHO director Margaret Chan at a regional summit in Conakry.

In Liberia all public buildings are disinfected.

Now Belgium also advises against all non-essential travel to Guinea, Sierra Leone, and Liberia.

Saturday, August 2, 2014

The US repatriates Kent Brantly. He is taken to the hospital of Emory University in Atlanta.

People from Ebola-stricken countries who still wish to attend the Africa summit in Washington DC will be screened medically before take-off and also upon arrival.

Airline company Emirates of the United Arab Emirates no longer flies to Guinea. This is the first intercontinental airline to cut its links with an Ebola-stricken country.

Monday, August 4, 2014

In Nigeria there is a second case of Ebola: Ameyo Adadevoh, a doctor from Lagos who cared for the Liberian lawyer Patrick Sawyer during the previous month. Seventy other people who possibly had contact with Sawyer are still being sought.

The World Bank announces emergency aid of up to 200 million dollars.

Tuesday, August 5, 2014

The American aid worker Nancy Writebol is also repatriated from Liberia to Atlanta in the US.

The current death toll of the epidemic in West Africa: 887 human lives.

Wednesday, August 6, 2014

It's going better with the repatriated American health workers Kent Brantly and Nancy Writebol after they received the experimental agent ZMapp,* even before they left Liberia. Doctor Brantly recovers more quickly than nurse Writebol.

ZMapp is made by Mapp Biopharmaceutical, a small private business in the United States. It is a serum mixture of three monoclonal antibodies that attack proteins on the surface of the Ebola virus. The WHO must still deliver its opinion on this type of experimental medicine. Another problem is the limited supply, giving rise to another pressing question: Who must receive the agent first?

African musicians have written a rap number about Ebola for the inhabitants of Guinea and Liberia. Music is an excellent method to spread a message among the many illiterate people in these lands. The song goes as follows: "Ebola, Ebola in town. Don't touch your friend! No kissing, no eating something. It's dangerous!" The "No eating" refers to the many miracle cures for Ebola that are offered to infected people. Concoctions of chocolate, coffee, sugar, and eggs, for example.

At first sight the song is a great initiative because sixty percent of the people in Guinea can't read or write. In Liberia it is forty percent. But I understand the critics who say that the lyrics encourage people to shun those who are infected, whereby they become socially isolated, even if they survive the disease. It is a double-edged sword. You must inform people, but then you make them afraid. You must physically isolate infected people and take the necessary preventive measures, but by doing so you brand people. And in every epidemic this happens again. Due to this, it is very important to start prioritizing basic health care right after an epidemic: clean water and affordable access to modest yet invaluable first line care, where people can also get the right information.

Spain repatriates Miguel Pajares, a 75-year-old cleric who was infected in Liberia. Father Pajares worked in Monrovia, in a hospital that has in the meantime been closed. While he was there among the patients, he cared for the hospital director, who has since died. By means of an Internet petition tens of thousands of Spaniards called on the government to repatriate the father.

In Geneva, WHO experts gather to consider new measures. It is possible that a kind of worldwide health emergency will be declared. Experts also plead for more experimental treatments.

Liberia declares a state of emergency for ninety days.

Thursday, August 7, 2014
In Nigeria doctors cancel strikes for better working conditions. The virus is in Lagos and its possible spread through the city's shanty towns would threaten to create a catastrophe without parallel. It is all hands on deck for the health workers.

In total there are now 932 deaths in West Africa. The United States has declared the highest health alarm. A steadily growing number of airlines adjusts their flights to West Africa or scraps them. Brussels Airlines is still flying but has arranged the flights such that personnel no longer need to stay overnight in the stricken countries.

Friday, August 8, 2014
The Nigerian president Goodluck Jonathan declares a state of emergency and reserves an extra 12 million dollars for isolation centers, detection, medical personnel, and border controls. Jonathan requests that no mass meetings are organized. Schools remain shut.

Sunday, August 10, 2014
In the north-west of the Democratic Republic of the Congo* Ebola is identified in a woman who ate bushmeat and who later dies. It is the seventh outbreak recorded in Congo.

The Spanish priest Miguel Pajares now receives ZMapp.

Tuesday, August 12, 2014
The 1,000 mark is passed, with 1,013 deaths out of 1,848 infections.

At 9:28 Miguel Pajares dies in the Madrid hospital La Pas—Carlo III. ZMapp hasn't helped. The manufacturer sends their whole supply to Liberia, free of charge.

Wednesday, August 13, 2014

In Nigeria, Jatto Asihu Abdulqudir, the official who helped Patrick Sawyer at the airport, dies.

Friday, August 15, 2014

Doctors Without Borders compares Ebola with a war and expects that the epidemic will certainly last for half a year. According to DWB the doctors cannot cope with the situation. "A new strategy is urgently needed, under the leadership of the WHO," announces DWB chairwoman Joanne Liu after a visit to West Africa. "If we can't stabilize Liberia, we won't succeed in stabilizing the whole region."

The WHO admits that they underestimated the epidemic.

Saturday, August 16, 2014

The World Food Program of the UN starts to provide food aid for more than one million people in Ebola areas.

In the Chinese city of Nanjing, the Olympic Games for Youth have started without three athletes from countries afflicted by Ebola. They are not welcome in China.

Sunday, August 17, 2014

In the Liberian capital Monrovia 17 patients have escaped from their quarantine in a hospital in the densely populated shanty town of West Point.

Monday, August 18, 2014

In the Damiaan Hospital in Oostende a youngster of 13 from Guinea is admitted with a high temperature. He has malaria, but perhaps also Ebola, and is kept in isolation.

Tuesday, August 19, 2014

New figures: 2,240 infections, of which 1,229 people have died. Liberia, the land hardest hit, introduces a night-time curfew. Two districts, including the shanty town West Point with 75,000 residents, are quarantined in their entirety.

Cameroon closes its border of 2,000 kilometers with Nigeria, even though the Ebola cases in Nigeria have all been diagnosed in Lagos, on the other side of the country.

There is relief in Oostende. The youngster from Guinea who is in the Damiaan Hospital does not have Ebola.

Wednesday, August 20, 2014

Doctors Without Borders speaks of a total catastrophe, especially because many infected people receive treatment that is far from optimal. If better treatment could be offered, thirty to fifty percent of the patients would survive the disease.

Liberian soldiers use tear gas against demonstrators and citizens who want to escape from quarantine at West Point. A young man dies of bullet wounds.

Thursday, August 21, 2014

The American mission doctor Kent Brantly is discharged from the Atlanta hospital. Nurse Nancy Writebol is also out of quarantine, but not yet the hospital. Brantly's blood will later be used to treat three other patients in the US.

Friday, August 22, 2014

In Nigeria for the first time there are secondary infections: partners of people who had contact with the Liberian-American lawyer Sawyer. Secondary infections are bad news for the most heavily populated country in Africa.

The World Health Organization works on a staged plan for the coming six to nine months, after sharp criticism from Doctors Without Borders, among others. Steadily more countries, including Cameroon, South Africa, Gabon, and Senegal close their borders to citizens from the stricken states.

Saturday, August 23, 2014

Sierra Leone institutes a new law: whoever hides a patient can go to prison for two years. No one knows the correct number of cases because many scared patients hide away, which only helps the Ebola virus.

There is a first British case, in Sierra Leone: William Pooley, a nurse of 29 years.

The Philippines withdraws its 115 peacekeeping troops from Liberia.

Sunday, August 24, 2014

Meanwhile in North Congo at least thirteen people have died from Ebola.

Great Britain evacuates nurse William Pooley from Sierra Leone.

In Kontich I receive a black letter in my postbox. Sister Genoveva is no more. Geno, my cheerful presenter at Yambuku Broadcasting System. Fortunately her last wish could be fulfilled, and Annie Ghysebrechts—so she was named at birth—is peacefully laid to rest, surrounded by her relatives.

Monday, August 25, 2014

Brussels Airlines no longer flies to Sierra Leone, Liberia, or Guinea, due to "operational reasons." How can aid workers and aid equipment still reach the Ebola countries?

Tuesday, August 26, 2014

Two Belgians speak out on the WHO: in the French newspaper *Libération* Peter Piot is quoted as saying that the World Health Organization "only woke up in July," while the epidemic had already begun in December 2013. Dr. Marleen Temmerman, gynecologist, well-known Belgian politician, and currently a departmental head at the WHO, admits that her organization realized the scale of the problem too late.

Brussels Airlines will once again fly to the Ebola countries, one reason being that the UN insists upon it. Ouch! The flight crews remain unhappy.

In Sierra Leone the WHO closes a lab after one of its staff members is infected.

Thursday, August 28, 2014

A doctor dies in the Nigerian city of Port Harcourt. It is the sixth fatality in this country, and the first time that the virus appears outside Lagos. The doctor treated someone who was in contact with the advocate Patrick Sawyer. His wife, also a doctor, is quarantined.

The top journal *Science* publishes an article about the genetic background of the virus in the current outbreak. By the time the article appears, five of its co-authors have died from Ebola.

Friday, August 29, 2014

A case is flagged in Senegal—a student from Guinea who wished to be treated in Dakar. He is immediately placed in isolation.

The president of Sierra Leone sacks his minister of health because he has not tackled the crisis well.

Saturday, August 30, 2014
The government of Liberia says that the quarantine in West Point, a district of the capital city Monrovia, which was in force from August 19, will be lifted at 18:00.

Sunday, August 31, 2014
In a test on 18 infected rhesus monkeys ZMapp appears to be 100 percent effective, reports the top journal *Nature*. Remarkably, some of the monkeys only received the medication five days after the beginning of the infection. Usually by this stage Ebola will kill a monkey, but now all the test animals survive it, including those treated late. During the present epidemic, many different people have already been treated with ZMapp. Some made it, others not. ZMapp has never been tested according to the rules for new drugs.

The wife of the doctor who died on August 28 in Nigeria, has also fallen ill.

In the largest four sea harbors of Liberia no sailors may henceforth board or disembark from ships.

Monday, September 1, 2014
The WHO's tally stands at roughly 3,000 cases with more than 1,500 deaths.

Tuesday, September 2, 2014
The British aid worker William Pooley is recovering well, his parents say. He is being treated in a London hospital with ZMapp.

In the American journal *PeerJ*, researchers from the University of Buffalo publish a study of fossilized rodents in which they posit that filoviridae, the family of viruses to which Ebola and Marburg belong, already existed 16 to 23 million years ago. Hence Ebola and Marburg would not have originated with the arrival of agriculture about 10,000 years ago, as is often accepted. They were already present in the Miocene and had a common ancestor. This is not only of scientific interest, but can also be important for the development of one vaccine against both Marburg and Ebola.

Wednesday, September 3, 2014

The epidemic in West Africa has already cost the lives of about 1,900 people, says Margaret Chan of the WHO, more than all previous Ebola outbreaks put together.

In Nigeria a seventh patient has died.

There is a third American patient. At the beginning of August, Dr. Rick Sacra departed for Liberia after his colleagues Brantly and Writebol were infected there. Among other things, he performed Cesarean sections on women with Ebola.

The British nurse William Pooley may leave the hospital, after blood is drawn from him for the treatment of future patients.

Thursday, September 4, 2014

Rick Sacra arrives in the US as the third American to be evacuated from Liberia. He is treated in Omaha, Nebraska, and will recover, thanks to a transfusion with blood from his recovered colleague Kent Brantly.

Saturday, September 6, 2014

The government of Sierra Leone announces a general national quarantine. Two weeks from now the inhabitants of Sierra Leone will not be allowed to leave their homes for four days. In West Africa 2,100 people have now died.

Tuesday, September 9, 2014

Roughly twenty Belgian doctors, among them virologist Marc Van Ranst, call on Belgium and other Western countries to send aid workers to West Africa, the B-FAST team, for example. They ask for a powerful signal from politicians, because the number of cases is rising exponentially. "Can we really wait until Ebola is knocking on our front door?" asks Van Ranst.

"The Belgian B-FAST team is improperly trained and equipped to be sent to West Africa now," reacts B-FAST doctor Nico Rooselaer. Foreign Affairs doesn't see this happening either. B-FAST serves for short and acute interventions, such as earthquakes and floods, and is not equipped for missions that last weeks, followed by a quarantine that lasts for a further 21 days.

During the coming weeks the WHO expects thousands of new infections in Liberia. There is a great shortage of beds and medical personnel, with the result that infected people who present themselves at hospitals are immediately sent away and can thus infect others.

A fourth American aid worker, a doctor who contracted the disease, is repatriated from Sierra Leone. He will come out on top thanks to the serum of the recovered Brit, William Pooley.

Friday, September 12, 2014

William Pooley says that he wants to return to Sierra Leone as quickly as possible. "The decision to return was not difficult," he says. "I am, after all, immune to this type of Ebola. I do not know how long the immunity* lasts and how watertight it is, but for me it is now in any case much safer than before to work there." At the beginning of October, William Pooley will indeed return to Freetown.

Two Dutch doctors still in Sierra Leone, Erdi Huizenga (a woman of 39) and Nick Zwinkels (31), are possibly infected. The previous week they were in contact with three patients at the Lion Hearts Medical Center in Yele—patients who later appeared to have Ebola—and they had not worn any protective clothing. Their patients had in the meantime died. The two Dutch doctors are repatriated and admitted to the University Hospital in Leiden for observation. They have no symptoms and test negative, but to make sure they do not infect anyone, they freely enter isolation at a secret location. The media fall on this story like a pack of wolves—the tropical doctors could not be thinking much of them.

Later, on September 30, the Lion Hearts Foundation makes it known that the two doctors are definitely not infected with Ebola. But they don't leave it there. At the end of October, Erdi Huizenga, against all advice and without letting anyone know, returns to Yele to resume her duties in the maternity clinic and continue the fight against Ebola. At the beginning of December Nick Zwinkels also returns to Sierra Leone. Respect!

Saturday, September 13, 2014

The number of Ebola deaths in West Africa now totals more than 2,400, out of roughly 4,800 infections.

Cuba sets a good example by sending 165 aid workers to Sierra Leone, of whom 62 are doctors.

Monday, September 15, 2014

We follow the tale of Thomas Eric Duncan in the Liberian capital city of Monrovia. This Monday his landlady, the pregnant Marthalene Williams, must go to the hospital. Nowhere does she find an ambulance. Fortunately, her tenant, the 42-year-old Thomas Eric, rushes to her aid. Eric fears a miscarriage and arranges a taxi, not knowing that Marthalene has Ebola. Together with her father and her brother, he accompanies Marthalene to the hospital. However, there is no space and thus they return home, where Eric helps to carry the woman out of the taxi. Marthalene dies not long after.

Tuesday, September 16, 2014

In Monrovia a French nurse of Doctors Without Borders is infected. This is the first French citizen to be infected in this outbreak. She will be taken to France. A good two weeks later it is declared that she is cured. On October 4 this aid worker may leave the military hospital of Bégin in Paris.

The United Nations needs 800 million euros to crush the epidemic. This is much more than was stated one month ago.

The United States is going to send 3,000 people to build health centers, to set up a coordination center, and to train health workers. President Obama calls the efforts to fight Ebola in West Africa "the greatest ever international reaction of the CDC." Obama says that the world is looking to the US to lead action in this crisis.

Thursday, September 18, 2014

Sierra Leone has instituted a three-day ban on outings from home, applicable to all residents, up to and including the weekend. During this time 21,000 health helpers will go from house to house to inform the population, to detect infections, and to distribute 1.5 million pieces of soap. DWB says that this measure makes no sense. For a door-to-door screening of people with Ebola symptoms you need experienced medical professionals. And if they do find infected people, there will be too few reception centers to care for them.

In Womey, south-east Guinea, the region where this epidemic began, village residents find eight bodies in the septic pit of a public latrine. They are directors, managers, government representatives, and journalists who were members of a public information team. Two days earlier the team had entered the region to make the people aware of the dangers of Ebola. Their public was not grateful. The delegation members were stoned by furious village residents and beaten with staffs. They still tried to flee but the majority of them were taken by force and later killed in cold blood, stated a journalist who managed to escape. The throats of three of the corpses in the latrine had been sliced through.

Part of the population in West Africa doesn't believe that the virus exists and blames health workers for the epidemic. In this case, the attack occurred not far from an area where a month earlier rows had broken out when a marketplace was disinfected with a chemical spray, and neighborhood residents had thought it was an attempt to infect them.

What happened in Womey isn't the first incident of its type and it also won't be the last. In the middle of February 2015 a transit center for Ebola patients in Faranah, Guinea, will be attacked and a DWB service vehicle set alight. In Guinea the Red Cross is on average attacked ten times per month during the course of 2014.

This makes me distraught and despondent. Friends of mine have suffered similar experiences during attempts to eradicate polio. Many aid workers sacrifice the luxury and security of office jobs to venture into the field and if need be they expose themselves to deadly diseases, because they now have a mission, because they are trying to save the lives of other people. How dreadfully sad it is when the intended recipients of this aid—through poverty, the clash of cultures, disinformation, and the incomprehensible deaths of village residents—have become so crazed by anxiety and hate and lack of understanding that they see no other solution than to attack their protectors, and in some cases even to take their lives.

The UN creates an independent unit, the United Nations Mission for Ebola Emergency Response (UNMEER).

Friday, September 19, 2014

Back to Thomas Eric Duncan in Liberia, who for the first time in his life leaves for America. Eric is unaware of the fact that he was infected with Ebola four days earlier when he helped his pregnant landlady. He has recently resigned from a job as a private chauffeur. He is eager to visit relatives in Dallas, Texas. One of Eric's sons, who he hasn't seen for a long time, now lives there with Eric's ex-wife and four other children of this woman. At the airport in Monrovia, Eric catches a Brussels Airlines flight to Belgium. Liberian officials—on the alert since the Patrick Sawyer incident—will later declare that on the questionnaire that passengers must complete, Eric lied about possible contact with the disease. In Brussels Eric boards flight 951 to Washington Dulles, where he transfers to flight 822 bound for Fort Worth, Dallas.

The UN Security Council holds an extraordinary session and approves a resolution about the epidemic in West Africa because "the unprecedented scale of the outbreak is a threat to international peace and security."

The UN has never before called an extraordinary session for a health crisis.

In Freetown, Sierra Leone, the streets are deserted.

Saturday, September 20, 2014
At 19:00 local time Eric Duncan arrives in Dallas. He stays overnight in the district of Fair Oaks, with his ex and her five children.

Sunday, September 21, 2014
Spain again evacuates an infected priest from Africa, Dr. Manuel García Viejo. This priest-doctor has been working for an NGO in Freetown for the past twelve years. He is the second patient Spain has repatriated. The previous one, Priest Miguel Pajares, is dead.

The other Ebola epidemic, in the north-west Congo, has already claimed 40 lives. There have been 71 suspected cases since the disease killed its first victim on August 11 in the region of Boende, Equatorial Province.

Switzerland repatriates a nurse who has been bitten by a child with Ebola in Sierra Leone. The man was wearing protective clothing and has no visible wound, but since opinions have been aired that in January there will be a global epidemic with one million deaths, the slightest risk is too great.

Monday, September 22, 2014
Nigeria now has 20 cases and 8 deaths.

In Monrovia, Liberia, a treatment center with 150 beds is opened. Patients flood in.

Tuesday, September 23, 2014
In Dallas, Eric Duncan doesn't feel well. He has a stomach ache.

Wednesday, September 24, 2014
Eric Duncan goes to the emergency service at the Texas Health Presbyterian Hospital in Dallas. According to the hospital, in the first instance it established that Eric had a mild fever (37.8°C/100.04°F), and Eric told them that he was suffering from stomach aches and a headache, and was urinating less than normal. The first aider asks questions about recent travel and notes that Eric has just arrived from Liberia but has not been in the vicinity of sick people. Duncan is sent home with the diagnosis "banal viral infection" and a prescription. From a medical file that is blown by a favorable wind to the offices of the Associated Press, it will later be seen that Eric was already running a temperature of 39°C (102.2°F) on that day and was suffering much pain.

Thursday, September 25, 2014
With immediate effect Sierra Leone places three regions and twelve areas, together containing over 1.2 million people, in quarantine. This is declared by President Ernest Bai Koroma on TV. How long this will last is unknown. Two districts were already placed in quarantine earlier.

In Madrid, Manual García Viejo, the Spaniard repatriated from Sierra Leone, dies.

Friday, September 26, 2014
The WHO announces that there have been more than 3,000 Ebola deaths and about 6,500 infections. At 1830 the number of deaths in Liberia is greatest, three times more than in neighboring Sierra Leone or Guinea, where the outbreak began. The WHO calls the epidemic the most serious public health emergency of modern times. Never before has Ebola claimed so many victims so quickly in such a vast area.

Also according to the WHO, by the beginning of 2015 thousands of doses of an Ebola vaccine must be available. This refers to vaccines made by the British firm GlaxoSmithKline (GSK) and the American NewLink Genetics (in cooperation with MSD). GSK should have 10,000 doses at hand, and NewLink Genetics a few thousand more. The stock of the experimental agent ZMapp has been exhausted, but before the end of 2014 several hundred new doses must come on the market.

Cuba sends 300 extra doctors to West Africa, which brings the number of Cuban doctors in the region to 465.

Sunday, September 28, 2014

In Dallas, Eric Duncan is vomiting. On this day he is once again taken to the emergency service at the Texas Health Presbyterian Hospital. There he must wait for approximately three hours. The nurses, among whom are Nina Pham and Amber Vinson, are exposed to his vomit and sweat.

Tuesday, September 30, 2014

The CDC confirms that in Dallas the Liberian Thomas Eric Duncan has Ebola.

In Senegal and Nigeria the situation appears to be under control, according to American and African health workers. In the past month no new cases have arisen in Nigeria; there the tally of infections remains at 20. In Senegal there has been only one new case.

Ivory Coast resumes flights to Liberia, out of humanitarian considerations.

Thursday, October 2, 2014

The United Nations World Food Program mentions food shortages and fears famine in parts of West Africa. Borders are closed, markets lie silent, and in some areas people must remain indoors. In Guinea, Liberia, and Sierra Leone it becomes ever harder for farmers to work their lands and collect harvests. Where there is food, its price rises due to scarcity. In Sierra Leone, a 61-year-old rice and manioc farmer who has six children and four grandchildren tells press agency Reuters that "Hunger will kill everyone who survives Ebola."

The official bill is now 3,338 deaths.

The Liberian president Ellen Johnson Sirleaf is furious and says that Thomas Eric Duncan can be prosecuted if he returns from the US to Liberia. "He has committed fraud in writing before his departure from Monrovia by stating on the form that he had not had any contact with an

Ebola patient, and in this manner he has harmed the inhabitants of a country that is doing so much in the fight against Ebola." Duncan himself says that he did not know that the woman he helped in Monrovia had Ebola.

Friday, October 3, 2014

Once again a patient is repatriated to the US: Ashoka Mukpo, a freelance cameraman who works for an American news service.

Saturday, October 4, 2014

The Dallas hospital lets it be known that the condition of Thomas Eric Duncan has worsened and is now critical.

According to the American health authorities, ten people in Texas are "at great risk" of having been infected by Duncan.

Monday, October 6, 2014

In Spain a 44-year-old assistant nurse, María Teresa Romero Ramos, is admitted to a Madrid hospital. This is the first case of an Ebola infection having occurred inside Spain, and also the first infection on European soil. This woman cared for the two Spanish priests who have already died. The Spanish government forms a crisis unit.

Tuesday, October 7, 2014

The US is going to set up three military labs in Liberia staffed by experts in biological warfare who will analyze the body fluids of infected patients.

Spain is scared. The husband and a colleague of María Teresa Romero Ramos must now enter quarantine. Roughly fifty people with whom she had contact are being monitored. According to some media, the hospital had been careless. It is alleged that the protective clothing did not meet the requirements of the WHO, and waste from the rooms of the Ebola patients was removed in an elevator for personnel. El Mundo reports that the nurse herself had several times asked for a test, and even after being tested still had to wait in casualty among other patients. The Spanish health authorities deny the media reports.

On American TV the Liberian ambassador says that his country is standing on the edge of a precipice.

Wednesday, October 8, 2014

The White House announces that passengers from the afflicted countries will be subjected to checks at five American international airports: JFK in New York, Newark, Chicago O'Hare, Washington Dulles, and Hartsfield-Jackson in Atlanta.

The British government sends 750 troops to Sierra Leone to help build a treatment center.

General-Major Geert Laire, chief of the medical service of the Belgian army, determines that a special center must be provided in the military hospital of Neder-over-Heembeek for a possible Ebola crisis in Belgium.

Thomas Eric Duncan dies.

Thursday, October 9, 2014

From now on passengers from Sierra Leone, Guinea, and Liberia will be checked at the British airports of Heathrow and Gatwick. Checks are also introduced on Eurostar, the train that connects London with Paris and Brussels.

Friday, October 10, 2014

There are now more than 4,000 registered deaths in the seven countries where Ebola has been reported.

The Flemish Agency for Care and Health (Het Vlaams Agentschap Zorg en Gezondheid) has drawn up guidelines for general practitioners and hospitals, which include a duty to report suspected cases and a special web page.

Ivory Coast bans the sale of bushmeat. Flesh from wild animals is generally to be regarded as a source of infection. Nigeria follows.

Excalibur, the dog of the infected Spanish nurse María Teresa Romero Ramos, is put to sleep as a precaution, in spite of a worldwide petition with 380,000 signatures. #SalvemosaExcalibur (let us save Excalibur) is now one of the world's most popular hashtags.

Saturday, October 11, 2014

In the Spanish capital Madrid, another two people who had contact with Teresa Romero are placed in quarantine. They are a nurse who cared for Romero and a hospital hairdresser. In Spain a total of sixteen people are now being monitored.

Nina Pham, the 29-year-old Vietnamese-American nurse who cared for Eric Duncan, has Ebola. Nina was wearing protective clothing, but it is possible that something went wrong when she was removing it. Apart from her, another 47 people who had contact with Duncan are being monitored.

I send Oyewale Tomori, professor at the Nigerian Academy of Sciences and an Ebola colleague from my earliest days, a congratulatory e-mail for the energetic manner in which his country has tackled Ebola. Although the virus is circulating in Lagos — with 12 million residents the biggest city in Africa after Cairo, a true epidemic has not arisen and there is a good chance that Nigeria will soon be declared Ebola-free.

Sunday, October 12, 2014

In Spain the condition of María Teresa Romero Ramos, who has been given ZMapp, is somewhat better.

Monday, October 13, 2014

The government of Liberia requests health workers not to participate in today's national strike. The National Union of Health workers, which called for the strike, is demanding a 200 euro increase in the risk premium. Health workers now receive 395 euros per month, in addition to a salary of 160 to 240 euros. They also want personal protective clothing and insurance for each Ebola worker. Roughly 200 Liberian health workers have been infected; 95 have died. The elections in Liberia have been postponed because gatherings of people at voting booths can aid the spread of the disease.

Tuesday, October 14, 2014

The Mayor of Dallas announces that Bentley, the one-year-old King Charles spaniel of the infected nurse Nina Pham, is being quarantined. Nurses who wish to remain anonymous state that at the Health Presbyterian Hospital in Dallas no specific safety protocols were followed during the treatment of Thomas Eric Duncan.

The staff of Brussels Airlines demands extra safety measures. Brussels Airlines is the only European airline company that still flies directly to and from Freetown in Sierra Leone, Monrovia in Liberia, and Conakry in Guinea.

Wednesday, October 15, 2014

In Texas the second nurse who cared for Thomas Eric Duncan, Amber Vinson, has contracted the virus. The CDC hunts for the passengers who one day earlier sat with Vinson on two different flights. The condition of her colleague Nina Pham improves.

Friday, October 17, 2014

Worried parents in Hazelhurst (Mississippi) take their children out of the local junior high after it becomes known that the school principal attended the funeral of a relative in Zambia, 4,500 kilometers south of the Ebola-stricken countries.

Dr. Erika Vlieghe, head of the tropical medicine service of Antwerp University Hospital (Universitair Ziekenhuis Antwerp), becomes the Belgian Ebola coordinator.

From now on baggage arriving at Brussels Airport from the Ebola countries is handled separately by a specialist business.

In an unpublished but leaked WHO document it is stated that the organization made systematic errors in the first phases of the epidemic.

Senegal is declared Ebola-free.

Saturday, October 18, 2014

On a Brussels Airlines flight to Washington DC a man from Liberia vomits. As a precaution Brussels Airlines flies the aircraft back to Belgium empty. In the meantime the man is tested. He doesn't have Ebola.

President Obama asks his people not to "surrender to hysteria." Two American nurses are now infected and one Liberian is dead, but there is no question of an epidemic in the US, Obama emphasizes.

Sunday, October 19, 2014

Nigeria, with 170 million inhabitants far and away the most populated land in Africa, is declared Ebola-free by the WHO after 42 days without new cases—that is twice the maximum incubation period. The first Ebola case was discovered on July 20 in the metropolis of Lagos. The government intervened rapidly, with the result of only 20 infections and 8 fatalities.

Belgian Ebola coordinator Vlieghe and newly inaugurated premier Charles Michel announce airport controls in Brussels. As from tomorrow, passengers from Ebola-stricken countries will be screened by a machine that stands between the aircraft parking bay and passport control. If they have a fever they will be examined by a nurse, and if need be by a doctor. The baggage handling will be stricter, too. Leaking baggage will be destroyed in Brussels and the other baggage disinfected. There will also be training for the safe cleaning of aircraft. The unions react with satisfaction.

From new tests it appears that the Spanish nurse Teresa Romero probably no longer has the virus.

The Health Presbyterian Hospital in Dallas concedes that errors were made in its handling of Thomas Eric Duncan. Nurses did not tell each other that Duncan had traveled to Africa shortly before, and there was no list of the (probable 125) people with whom he came into contact before he died.

On the BBC the Liberian president Ellen Johnson Sirleaf, who won the Nobel Prize for Peace in 2011, asks the international community for tougher action because "the time for talking and theorizing has passed." According to her, alongside the humanitarian disaster, an economic one is threatening to break out, due to all the missed harvests and the closed markets and borders. After the end of the civil war in 2003, this land began to recover, but now it has again been knocked down. In Liberia some 2,200 people have succumbed.

Monday, October 20, 2014
Doctors Without Borders in Belgium starts a campaign to gather funds. The organization believes that the government is doing too little.

Tuesday, October 21, 2014
The Belgian minister of Development Cooperation Alexander De Croo makes seven million euros available for the fight against Ebola in West Africa. The money goes to UNICEF, the Red Cross, and Doctors Without Borders.

According to *The Lancet* it is cheaper and better to check travelers in the departure airports of West Africa than in the arrival airports. In the three international airports of Guinea, Sierra Leone, and Liberia this is indeed happening. Controls at airports outside Africa do create a feeling of security, but deliver a minimal benefit. That money can be better spent on other health measures. Nevertheless, France, the United Kingdom, and

Belgium check passengers arriving from the afflicted countries, and the United States checks the temperatures of passengers in some aircraft.

Sometimes I ask myself if politicians should not look at the science a bit more before they make decisions. Would they know what has appeared in *The Lancet*? Perhaps the article hasn't got it right—that is possible—but *The Lancet* is definitely the best source of information. Do politicians do what their voters want? Or do they do what is best for everyone according to science?

Wednesday, October 22, 2014

The Spanish nurse Teresa Romero is officially cured.

Also the freelance cameraman, who was the fifth person to be repatriated to the US, has recovered.

Bentley, Nina Pham's little dog, does not have Ebola. Hallelujah!

Thursday, October 23, 2014

Ebola now pops up in Mali, the sixth West African country to be stricken. The victim is a two-year-old girl who was taken on a visit to Guinea, and when back in Mali, is brought to a hospital in Kayes, where she tests positive. According to the government the people with whom the toddler had contact in Mali have been identified and are under medical supervision.

Under the leadership of the Institute for Tropical Medicine in Antwerp, the EU, the WHO, and the research network ISARIC are going to support research into the treatment of infected patients with antibodies from the blood of recovered patients. The investigation is taking place in Guinea. It is not the first time that this treatment has been used; in the Zairean outbreak at Kikwit in 1995, blood and plasma therapy were applied successfully, albeit without preceding clinical studies. We have also studied it for HIV, but without substantive results. The method has been tested for certain other diseases and found to be safe. According to the project leader the therapy can also help to reintegrate recovered patients—as donors of the therapeutic medication—into the community once again. Now recovered people are often looked at askance or even shunned.

The WHO holds a short-order consultation with scientists, governmental institutions, Doctors Without Borders, drug agencies, financiers, and representatives of the pharmaceutical industry. Access to experimental Ebola vaccines is on the agenda. The most important conclusions: the necessary funds must and will be provided; in the case of legal cases arising from the development of Ebola vaccines, no single country or pharmaceutical business will have to bear the costs of indemnification on its own, but will be backed by a club of donors and the World Bank; and the pharmaceutical companies confirm that there will be enough vaccine doses. GlaxoSmithKline can increase its monthly production capacity tenfold by April 2015 to 230,000 doses. NewLink Genetics and MSD can boost the production of their vaccine from 52,000 to 5.2 million, regardless of the chosen dosage. In the administration of vaccines, priority will be given to health workers such as doctors, nurses, and laboratory technicians, and also to burial and decontamination teams.

Craig Spencer, a 33-year-old doctor from New York who traveled back to the US from Guinea on October 14–17, appears to be infected and is placed in quarantine at the Bellevue Hospital in New York City. In Africa he treated Ebola patients for Doctors Without Borders. During his trip he had no symptoms, but as a precaution he took his temperature twice per day and on October 22, when he had already been back in The Big Apple for five days, he developed a fever. After his arrival in the US, Craig took the metro and an Ubertaxi and has been at a bowling alley. The three people with whom Craig has been in contact in New York are still healthy but are being monitored.

Friday, October 24, 2014

The two-year-old girl in Mali has died. She was the first confirmed case in this land. The WHO fears that another 43 people in Mali have been exposed to the virus.

The two nurses in Dallas, Nina Pham and Amber Vinson, no longer have Ebola. Nina Pham is invited to the White House by President Obama and is received there with open arms.

The governors of the American states of New York and New Jersey decide that medical personnel returning from an Ebola-stricken country must remain in quarantine for three weeks, a measure that is stricter than that requested by the CDC. It is immediately enforced: nurse Kaci Hickox is just back from Sierra Leone, where she helped Ebola patients. Upon her arrival at Newark, she is instantly placed in quarantine.

The retiring chairman of the European Council Van Rompuy tweets that the EU is increasing its support for the fight against Ebola in West Africa to 1 billion euros. The member states had already promised 600 million euros.

Reading this report finally gives me some hope. Clearly the international community has gradually realized that money is required to halt Ebola—lots of money. Hopefully international money donors such as the EU, the World Bank, and the International Monetary Fund will monitor the funds scrupulously to ensure that their dollars are used for the right purposes. And that reserves are established! Only then will there be enough money for basic health care infrastructure after the epidemic is extinguished. Only then will the events that followed the Yambuku outbreak not recur in West Africa. This support is needed not only against Ebola, but also against AIDS, TB, malaria, cancer, respiratory ailments, and against so-called First World diseases such as coronary artery disease, diabetes, and obesity, which are increasing in poor countries hand over fist.

Saturday, October 25, 2014
In New York there is a demonstration against discrimination. One demonstrator: "Even here in New York people avoid you when you are wearing African dress and you take the train or go shopping."

Mauritania closes its border with Mali.

Sunday, October 26, 2014
The state of Illinois also requires people who have had contact with Ebola patients in West Africa to remain in quarantine for 21 days after entering the US.

Belgium donates an extra 2.5 million euros for air transport. In this way the United Nations can deliver medical and protective supplies to the stricken countries. The money is in addition to the 7 million euros given to Doctors Without Borders, UNICEF, and the International Red Cross.

Monday, October 27, 2014
In West Africa more than 5,000 people have now died from the virus.

Australia applies visa restrictions to travelers from Ebola-stricken countries.

All American soldiers returning from Liberia and Senegal must spend three weeks in quarantine at the American airforce base near the Italian town of Vicenza. Right now this applies to twelve soldiers, who have no symptoms. In Liberia there are 600 American soldiers, in Senegal 100.

UN secretary-general Ban Ki-Moon is highly critical of the measure to automatically quarantine returning health workers and finds it unjust that people who risk their lives to fight the virus encounter problems and can become discouraged.

It appears that since mid-August Belgium no longer undertakes the enforced repatriation of asylum-seekers to Sierra Leone, Liberia, and Guinea, because the situation there is too unsafe. In the Netherlands a judge who had earlier decided to repatriate a Liberian "because Ebola is no reason not to do this," reverses his decision.

The government allows the ITM in Antwerp to undertake Ebola tests in its own laboratory. Because the tests no longer need to take place in foreign countries, a negative result can be given within four hours after the arrival of a suspect blood sample, says Professor Ariën, my successor as the head of virology at the ITM. Until now the samples had to be sent to a BSL-4 lab in the German city of Hamburg. The ITM only has a BSL-3 lab, but within it a separate space has now been created where extra security measures apply.

Tuesday, October 28, 2014

According to the new guidelines of the American Institute for Public Health, medical personnel who are returning from Ebola-stricken countries need not enter quarantine, regardless of the measures that some states have taken. The governor of New Jersey defends the quarantine of nurse Kaci Hickox and says that it remains in force. She can spend the rest of her quarantine at home. The affair moves the US enormously.

Wednesday, October 29, 2014

According to the WHO it looks like the number of new infections in Liberia is diminishing. There are fewer burials, and at these burials—this is still a substantial number—the traditional rites are being foregone.

Worldwide to date there have been 13,700 cases and an unknown but probably large number of unregistered cases.

Stephen Opayemi, a Nigerian resident of the US who attended a wedding in Nigeria at the beginning of October with his daughter, lodges a legal case against the management of the school where his seven-year-old daughter Ikeoluwa is being taught. After the journey Ikeoluwa was not allowed to enter the school. The teaching authorities and other parents were afraid that the child was infected. Nigeria had only been declared Ebola-free on October 19, but upon her return the girl did not have a single symptom. Opayemi believes that his daughter is being subjected to discrimination out of "anxiety and ignorance."

Thursday, October 30, 2014

From now on all foreigners who wish to visit North Korea must spend 21 days in quarantine, regardless of where they have come from. A remarkable measure for a closed country where there has not been a single Ebola case or suspected infection, and which very few foreigners are eager to enter.

Friday, October 31, 2014

Canada no longer issues visas to the citizens of Ebola-stricken countries or people who have stayed there for the past three months. Canada is the second western country to do this, after Australia.

In the US there is once again a rumpus after nurse Kaci Hickox, despite her compulsory quarantine, goes out for a short bicycle ride. The judge decides that she must remain subject to the mandatory health regulations, but that she may move freely provided that she displays no symptoms. "People react out of anxiety, and that is not always reasonable," declares the judge.

Sunday, November 2, 2014

In the United States the Associated Press and other press organizations announce that from now on they will only report on suspected cases inside the country if a test appears to be positive.

Tuesday, November 4, 2014

On the Flemish TV program *Reyers Laat*, two young aid workers of Doctors Without Borders who have just returned from a mission give firsthand accounts. Nurse Jolien Colpaert: "When I lifted a girl who was not even one-year-old from her cot, I felt that she was extremely hot. After giving her milk for five minutes, she died in my arms." Rob D'Hondt, who checked safety protocols: "The bodies of the dead patients are highly infectious. The number of viruses increases in step with how sick you are. When you die, that number is at its highest level."

Wednesday, November 5, 2014

The American president Barack Obama asks Congress to make an extra 6 billion dollars available. Three quarters of that money is required for immediate measures in the US and West Africa. The rest must be placed in a fund to cover unpredictable developments in the epidemic.

Thursday, November 6, 2014

Facebook launches an app where money can be donated and a hundred communication satellites steered toward the Ebola countries to support the Internet and telephone networks there. In this way, foreign aid workers and patients in quarantine can more easily stay in contact with family and friends. Facebook boss Mark Zuckerberg and his wife have donated 25 million dollars.

Journalist Katrien van der Schoot, from the Flemish Vlaamse Radio- en Televisieomroep station (VRT), reports while on her way to West Africa: "It is a scandal that the real help is only now coming on line, after panic in the US and Europe. The media has so skewed the picture that Ebola in Sierra Leone, for example, almost appears less in the news than the quarantine question in the US."

France is going to systematically screen passengers from Guinea.

Friday, November 7, 2014

Three robotics centers meet in the White House to discuss the possible use of teleguided robots to further restrict contact between aid workers and the virus. It is foreseen that robots could help with the removal of protective clothing and at funerals.

According to DWB the number of cases in Liberia is dropping rapidly.

Monday, November 10, 2014

Dr. Craig Spencer no longer has Ebola and may leave the Bellevue Hospital in Manhattan. In the United States there are no more known cases of Ebola.

The Irish singer Bob Geldof makes it known that he is going to organize a new benefit event for the fight against Ebola. There will be a fourth version of "Do they know it's Christmas?" Among others, One Direction, Ed Sheeran, Bono, Sinèad O'Connor, and Adele are already working on it. David Bowie will also be asked to contribute. In 1984 the first version was recorded. Then, the profits went to relieve famine in Ethiopia. The initiative was repeated in 1989 and 2004. The newest version of the single is for sale from November 17.

Teresa Romero, the Spanish assistant nurse who has recovered from Ebola, claims 150,000 euros in compensation from the Spanish government for her dog Excalibur. As a precautionary measure the dog was destroyed after Romero was infected. She believes that Excalibur had "been given no chance" because he was never tested for Ebola. According to usually well-informed sources, Romero is planning to buy a new puppy.

Tuesday, November 11, 2014

There are two new deaths in Mali: a patient from Guinea and the nurse who was treating her. The hospital in the capital city Bamako where she worked has in the meantime been shut. The first Ebola death in Mali was a child of two. The new cases are unconnected. Mali thought that it had escaped from the epidemic largely unscathed, but that has yet to be demonstrated.

According to Flemish pilots our harbors are far from watertight against possible infection. The pilots are also afraid of becoming infected themselves. Almost every day ships from West Africa moor here. The pilots say that in their work they stand in close physical contact with crews for a minimum of eight hours per day and that they often receive offers of a meal and drinks. It must be jolly.

Wednesday, November 12, 2014

Regardless of previous policy, the Belgian fast reaction team B-FAST now swings into action and will depart for Guinea with its light, mobile, and rapidly assembled B-Life-lab, as soon as the French army can guarantee security there. The team of eight people will stay there for two months. There were long discussions about the intervention of B-FAST.

Thursday, November 13, 2014

Liberia lifts the state of emergency that has been enforced since August. The number of new cases in that country is decreasing. The nightly curfew nonetheless remains in place.

Ebola has now cost 5,160 human lives out of a total 14,098 infections. In Guinea and Liberia it seems that the disease is now waning, but in Sierra Leone it is still increasing.

In my mailbox I find a message from Jim Lange, the man who supervised me in 1977 when I trained at the maximum security laboratories of the CDC in Atlanta. He mails: "Good morning old friend, I am greeting you from Liberia, where I am fighting Ebola with my colleagues from the CDC." Jim had in the meantime become a member of a team formed to combat bioterrorism, but is working temporarily in Liberia to monitor how well people who have been in contact with Ebola patients are tracked down. He searches for infected people himself, often in remote places that are difficult to reach.

Now that I again have contact with Jim, we e-mail to and fro. Apparently Jim bumped into Stan Foster a short while ago in Atlanta. He gives me Stan's e-mail address. And so I once again start chatting with my gracious host of former times. It was wonderful to stay with his family, and when Stan writes to me that after three months I will always be a part of the Foster household, I feel just as pampered and grateful as I was back then. To "foster" means to "nourish, cherish, accept into a family," says the Van Dale dictionary. That I can endorse.

The new outbreak also brings me into contact with Joel Breman of the CDC, once again. Joel is writing an abstract* of an article about Ebola and makes me a friendly offer to be a co-author. The final manuscript will be accepted for publication in May 2016 by The Journal of Infectious Diseases under the title: "Discovery and Description of Ebola Zaire Virus 1976, and Relevance to the West African Epidemic, 2013-2016." As a co-author I find myself in the company of Joel G. Breman, David L. Heymann, Graham Lloyd, Joseph B. McCormick, Malonga Miatudila, Frederick A. Murphy, Jean-Jacques Muyembé-Tamfun, Peter Piot, Jean-François Ruppol, Pierre Sureau, and Karl M. Johnson - with a couple of additions, the fine flower of 1976. This paper is dedicated to the devoted health-workers at the Yambuku Mission Hospital who first confronted Ebola in 1976, and many of whom succumbed to the disease. And so Ebola remains the cement of my life's mosaic.

Saturday, November 15, 2014

A Sierra Leonean doctor who lives in the US is infected in Freetown and is flown to the Nebraska Medical Center. He is sicker than the other people who have thus far been treated in the US.

Monday, November 17, 2014

The doctor from Sierra Leone who is being cared for in Nebraska dies.

Wednesday, November 19, 2014

Is Ebola getting its feet on the ground in Mali? There is now an eighth fatality. In a hospital in Bamako a doctor who treated a patient from Guinea has died.

According to the World Bank, almost half of the active population of Liberia is unemployed due to the epidemic. Those who are self-supporting are hardest hit.

The new official account of the WHO: 5,420 deaths and 15,145 infections in eight different countries.

Friday, November 21, 2014

The WHO announces that the Ebola epidemic in the Congolese Equatorial Province is officially over. The outbreak there, it would appear, is unconnected with that in West Africa. Sixty-six people were infected, of whom 49 died.

Singer Carla Bruni has written a French version of Band Aid 30's "Do they know it's Christmas?," Bob Geldof's initiative in support of the fight against Ebola. "Noël est là" is sung by French artists such as Vanessa Paradis, Joey Star, Zaz, and Bruni herself. Bruni's husband, ex-president Nicolas Sarkozy, also participates. Despite criticism of the project—it is alleged that the lyrics are dull and Christmas is incompatible with the culture of many people in West Africa—the English language version enters the hit lists of more than fifty countries and raises about 1 million euros immediately after its launch. A German version also exists.

Thursday, November 27, 2014

In the US the pharmacy company GlaxoSmithKline has tested an experimental vaccine on twenty healthy volunteers. Large scale testing of the vaccine should start in Liberia at the beginning of 2015. GSK says that by the end of 2015 it will be able to produce one million doses per month.

Friday, November 28, 2014

The French president François Hollande travels to Guinea with "a message of solidarity." He will visit hospitals there. Hollande is the first western politician to travel to the African country since Ebola broke out in March.

Thursday, December 4, 2014

A Dutch campaign to aid Ebola-stricken areas in West Africa has raised approximately nine million euros in one month. Among other things, aid organizations send twelve containers with medical supplies such as masks and protective goggles and clothing. Four pick-ups, ten motorbikes, and one hundred and fifty bicycles are also going, for the transport of the sick and the dead, but also for health care personnel and health educators who visit villagers to inform them.

According to the latest figures, the virus has claimed the lives of more than 6,000 people to date. Positive news: the WHO estimates that 70 percent of funerals are now being undertaken in responsible ways.

Saturday, December 6, 2014

An infected Nigerian soldier of the UN peacekeeping force in Liberia arrives at Schiphol Airport. He is cared for in a hospital in Utrecht. He is the first Ebola patient to be treated in the Low Countries, exactly a year after the beginning of this outbreak. There have already been a couple of Ebola alarms in Belgium and the Netherlands, but the virus was never confirmed. This man will recover and on December 19 it will be announced that he has been cured.

Wednesday, December 10, 2014

The American newspaper *Time* declares the doctors, nurses, and other people who are fighting Ebola to be "Person of the Year." About two hundred health workers have already died during their war against the virus in West Africa.

Bravo, *Time*. It is indeed thanks to the enormous dedication of health workers such as Doctors Without Borders that Ebola patients are being treated and that infection in hospitals is being averted, so that the trust of people in health care institutions is renewed. The latter, especially, is very important in order to get a grip on the epidemic. In the beginning, many people fled from health care centers or didn't want to enter them. Now, when they come out of them, they know that they are cured or know that they are not infected with the virus and instead have malaria, for example, which can be treated. The health workers have also seen to it that the deceased are buried in a safe and fitting manner.

However, *Time*'s acclamation comes too late for the many health workers who have themselves fallen prey to the disease. Certainly at the start of the epidemic, medical personnel everywhere were not fully informed and people didn't always work precisely and safely. Often the circumstances simply didn't allow the risk of infection to be eliminated: too few staff, bad protective clothing, too few transport options to move patients... There are no exact figures, but it is probable that roughly five percent of the cases are health workers, or at this time about 850 people. This is not only really serious for the doctors and nurses themselves, but also for current and future Ebola patients, and for West African patients in general.

Hence the attack on the AIDS epidemic in the region was sidelined. The reduction in HIV tests, the treatment of people with HIV and AIDS... all that is thoroughly disrupted now that health workers have their hands full with Ebola or have even become its victims. Even before the outbreak there was a shortage of trained personnel. The stigma attached to Ebola in the meantime also strikes people with HIV and AIDS. It throws them back to the beginning of their illness and the shame that went along with it. Some of them dare not enter their clinics or treatment centers any more. And this is true not only for HIV but also many other widely distributed diseases in the region. So it is that the Ebola epidemic hinders health workers intervening efficiently in the cholera outbreak that has been present for three to five years in the three West African countries.

Friday, December 12, 2014

In Sierra Leone this year, Christmas and New Year may not be celebrated, at least not in public. Although it is mostly an Islamic country, almost everyone in the land celebrates Christmas, and usually they do not do so in the isolation of their houses, but outside in the streets.

Saturday, December 20, 2014

In the morning a Belgian military B-FAST team leaves for Guinea to install a mobile laboratory there. The departure was already announced in November. The team of doctors and laboratory workers will stay in Guinea for two months. It will also test an experimental Japanese medication, the efficacy of which has already been demonstrated against the flu and other illnesses. The soldiers have been trained how to protect themselves and in principle will not come into contact with patients.

According to new figures, which are still climbing, more than 19,000 people have now been infected, of whom 7,400 have died.

Sunday, December 28, 2014

Pauline Cafferkey, a 39-year-old Scottish woman who was in the fray as a health worker for the NGO Save the Children in the Sierra Leonean center of Kerry Town, falls ill after returning to Britain.

Monday, December 29, 2014

In Glasgow, Pauline tests positive for Ebola. She is placed in isolation. The Scottish government holds urgent consultations and first minister Nicola Sturgeon lets it be known that the government is "well prepared." This is the second case of Ebola in the United Kingdom, after William Pooley.

Tuesday, December 30, 2014

Pauline is transferred to the Royal Free Hospital in London. She receives experimental antiviral treatment and blood plasma from someone who died.

Wednesday, December 31, 2014

On New Year's Eve, Peter Verlinden, Africa specialist of the Flemish VRT radio and TV broadcasting station, provides a synopsis of Ebola in the year 2014 in his blog on *deredactie.be*. The year had actually already begun in December 2013, with the death of the two-year-old Emile Ouamouno in the Guinean village Meliandou, but it would take until the fall of 2014 before the world began to worry about it, writes Verlinden. The first hundred deaths took place in silence. Yet Doctors Without Borders had already rung the alarm bell before summer. Only when the tally stood at three thousand at the end of September did the media send their reporters.

But this epidemic only really became world news after Ebola cases appeared in the US and Spain—people repatriated from Ebola areas and people who cared for those repatriated to western lands. More ink, bandwidth, and video time would be spent on them than on West Africa itself, although each month there are thousands of new cases there, Verlinden notes. The media that went to the trouble to send reporters to the spot has in the meantime recalled them. Not because the epidemic is over, but because the novelty has worn off. Yes, charity begins at home, but under these circumstances such a policy is an abuse. Also the Ebola victims who reach the rich northern countries receive the best possible Ebola treatment and often survive, while the Africans drop like flies on their own continent. In Flanders, too, we are more concerned with the risk of importing the illness than with our African fellow man. "But," questions Verlinden with justification, "is a (possible) Ebola infection in our own country then truly so much more important than the almost 20,000 infections and 8,000 deaths in West Africa to date? (...) Moreover, malaria, together with HIV/AIDS, is the deadliest African disease, with more deaths in the whole of Africa in one day than this Ebola epidemic has inflicted over a whole year up to the present. But that is another story."

His overview is painfully clear and in a first reflex I completely share Peter Verlinden's indignation concerning the spectacle-sick media and narrow-minded public opinion. Africa only becomes a problem when westerners share its misfortunes. It has never been otherwise. Sometimes Africans even play this shocking little game themselves. In 1976 we only heard about Yambuku when Flemish nuns died... Lassa? The same story. Western media are stubbornly uninterested in African problems and at best they wait until they get a western lead, if one should ever arise.

Yet Africa-lovers such as Peter Verlinden and yours truly desperately need the same media when we wish to move public opinion and politicians to consider expensive answers to public health problems, wherever they may be in the world. And to pay—and therefore asking them not to spend their budgets on other things. One euro that the government spends on the prevention of suicide, cycle paths, my pension, or the VRT broadcasting station can no longer be spent on an Ebola campaign in Sierra Leone. It is always a question of setting priorities.

But I continue thinking this over. Yes, we can make the citizen or media feel guilty, but even if that is successful, what do we achieve by doing so? Many Flemings naturally worry more about the risk—however small—that a terrifying, extremely deadly disease such as Ebola will break out in their own land than about the fate of unknown poor devils thousands of kilometers away. We are not the giant Atlas. One cannot possibly bear the weight of the world on one's shoulders without being crushed. In the first place people find themselves, their partner, their children, parents, and other loved ones important. Only thereafter the rest of the world, in steadily expanding concentric circles of decreasing affinity. It is hard but logical that the media knows this and acts accordingly, as an extension of our own interest, otherwise they would price themselves out of the market. News is what interests people, not what the planet truly needs.

If Peter Verlinden and I are especially preoccupied with Africa, that is probably more the result of accidents during the course of a human life and personal interest than because we are wiser or more humane than the average citizen. I do not wish to be accused of pretention. While Verlinden and I write about Ebola and are passionately concerned about seeing the great wicked forest through the trees of fear, to an equal extent we neglect the lot of alcohol-addicted native Americans or the underwater clouds of plastic particles that are choking the seabirds of the world's oceans—to name just a couple unsolved problems of the world. Not to mention the countless traffic victims, the growing resistance of diseases to antibiotics, drug addiction, religious fundamentalism, civil wars, famines, lack of water, climate change, racism, leaking nuclear power stations, deforestation, disappearing animal species, space junk, overpopulation in one continent, and top-heavy population pyramids in another. Cherokees with liver disease or gagging albatrosses only interest us when we read something fascinating or see something spectacular on TV.

I could decide that henceforth I will find all these concerns just as repulsive as the scientists and journalists who are indignant about them, but in this way the list of suffering would grow too long for me. I simply couldn't cope. And I don't need to find all of it in my daily newspaper, because then I would not be able to choose what I care about most.

Precisely because charity starts at home, life remains bearable despite all the misery, and now and then it is even joyful, and a person preserves enough energy to focus on what he or she just happens to find important. Let us scientists and journalists do our damnedest to find and check the facts, expose the underlying mechanisms, and so inform our readers, spectators, and politicians as well and as broadly as possibly, without targeting them with hopeless reproaches. In the end, perhaps that helps a little to determine the order of priorities.

Saturday, January 3, 2015
Pauline's condition is critical, her doctors announce in London.

Sunday, January 4, 2015
In the Nebraska Medical Center an American nurse who became infected in Sierra Leone is transferred from care to observation. Three patients have already received care in this hospital. Two of them survived the virus, but a doctor from Sierra Leone died there in November.

Monday, January 5, 2015
Pauline's condition is still critical, but stable.

Tuesday, January 6, 2015
Today the American pharmaceutical company Johnson & Johnson begins to test the Ebola vaccine of its Belgian subsidiary Janssen on 72 British volunteers. This vaccine combines the vaccine of Janssen subsidiary Crucell with that of the Danish firm Bavarian Nordic. The volunteers receive two doses: the first must prime the immune system, the second must strengthen the defense of the immune system. The test will last a year. Experiments on monkeys have shown that the vaccine offers full protection against Ebola. Johnson & Johnson accelerated the research because so many people in West Africa are still suf-

fering. In the next phase, which starts in three months, the pharmaceutical company wants to test the vaccine on thousands of people in Africa and Europe.

If things go well the vaccine will be usable in the afflicted countries by June or July. This year Johnson & Johnson want to make another two million doses of the vaccine. GSK and NewLink Genetics (together with MSD) are also working on a vaccine against Ebola.

For all sorts of reasons it would be fantastic if we had an effective preventive vaccine, even if it could only be used on a limited scale. If just the health workers could be vaccinated before it becomes more widely available, they could care for people with Ebola without being hindered by uncomfortable protective clothing that makes them rapidly become overheated. In this way, aid workers will also frighten patients less, communicate with them more easily, and in general build greater trust.

More than 8,000 people have now died and at least 20,691 have been infected, announces the WHO.

Wednesday, January 7, 2015
Al Qaeda has two religious fanatics murder the editors of the satirical paper Charlie Hebdo. Twelve people die, among which are five cartoonists. The rest of the day I do not think about Ebola and set myself to sketching, so long as this is still possible.

Thursday, January 8, 2015
The German Primate Center has developed a suitcase-sized mobile unit with which aid workers in remote areas can tell within quarter of an hour if a patient is infected. This is six to ten times faster than the current detection methods. The unit has a solar panel and will be tested in Ebola centers in Guinea. In the 70s and 80s we still had to lug around extremely heavy generators.

Sunday, January 18, 2015
In Mali the epidemic is over. All told, there were eight infections in this country. Six of the eight patients died. The cases followed each other in short order, which indicated that a new branch of the epidemic was in the making, but this has not happened.

Monday, January 19, 2015

In Guinea the schools are once again open. The classrooms are disinfected and the pupils are checked for fever. Of course, children with symptoms are not welcome. Nevertheless, many parents remain frightened. It will still be quite a while before teaching is back up to speed.

Tuesday, January 20, 2015

According to a new World Bank report, the economic impact of the epidemic on the whole of sub-Saharan Africa won't amount to 21.6 billion euros, as was still being stated in October 2014, but in the worst case 5.4 billion euros. That is because there is now less chance that Ebola will rampage through countries outside the three that have been stricken. For Guinea, Sierra Leone, and Liberia, the economic damage in 2015 will amount to a cool 2 billion euros.

According to the WHO, the Ebola epidemic has cost the lives of almost 8,500 people.

Wednesday, January 21, 2015

Guinea, Sierra Leone, and Liberia announce the lowest number of infections per week since August 2014.

Thursday, January 22, 2015

In Sierra Leone all quarantines are lifted now that the transmission of Ebola has drastically decreased.

Friday, January 23, 2015

The first three hundred flagons of the experimental Ebola vaccine developed by the pharmaceutical company GlaxoSmithKline in the Wallonian town of Rixensart are flown from Brussels Airport to Liberia. There the experimental vaccine will be tested on thirty thousand people. Whether the agent will actually be launched on the market, according to GSK, will depend on the degree to which the vaccine offers protection against Ebola without causing serious side-effects.

Saturday, January 24, 2015

Liberia announces just 21 probable and 5 confirmed new cases in the whole country for the past week. The epidemic is probably on its last legs. At the epidemic's peak in Liberia during August and September 2014, there were more than 300 new cases each week.

Pauline Cafferkey, the nurse who was infected in Sierra Leone and who fell ill during her journey back to her home in Scotland, has now fully recovered. At one time it was feared that she might die, but now she may leave the London hospital.

After a month in the Guinean town of N'Zérékoré, the Belgian lab B-Life has analyzed 190 specimens. The nurses and doctors take blood and saliva samples daily. Occasionally they also take samples of urine, mother's milk, and sweat to learn more about the evolution of the disease. For the first team the job is over. Today a second team arrives to relieve them.

Sunday, January 25, 2015
The latest WHO figures: 22,100 cases and 8,810 deaths.

Wednesday, January 26, 2015
After five months Senegal has once again opened its border with Guinea.

Tuesday, January 27, 2015
Oxfam wants a Marshall Plan for Sierra Leone, Guinea, and Liberia, where an economic catastrophe is looming. Just as Europe was helped by the US after World War II, so too must the Ebola-devastated lands now be aided by the rich countries, with money and measures to generate work opportunities, and to boost teaching and health care. In Liberia, for example, the price of food has risen by forty percent. "The world reacted to the Ebola crisis late and cannot offer any excuse as to why it shouldn't help the population and the economy," declares Oxfam.

Monday, February 9, 2015
Preventive campaigns for the safe burial of Ebola victims and others are facing great difficulties. According to the WHO unsafe burials are still occurring in the Ebola-stricken countries. Aid workers in certain districts are still encountering aggression from local residents and a lack of will to report suspected patients to the government. People are afraid that they may no longer approach their dead relatives and bid them farewell in the centuries-old traditional manner. So, of course, there is a chance that we will never be able to find the last EBOV infected person

(EBOV-zero) and that the infection will continue to slumber in Guinea, Liberia, and Sierra Leone, only to return in full force at some moment.

Tuesday, February 10, 2015
The Belgian company Biocartis in Mechelen is going to launch a new test on the market, the Rapid Ebola Virus Triage test, with which both EBOV and SUDV can be genetically detected in a single sample in one and a half hours. The test was developed in collaboration with the Virology unit of the ITM under the leadership of Kevin Ariën. EBOV and SUDV are two different species of Ebola viruses. With the test, aid workers can investigate blood and other body fluids, cells, or fixed tissue material, even in primitive field conditions.

Wednesday, February 11, 2015
The United Nations cautiously forecasts that we will arrive at the magical "Ebola zero" moment in the three West African countries at some time in June 2015. I have my doubts about this for two reasons: first, dangerous funeral rituals persist, and second, the worrying fact that for the first time this year there has been a rise in the number of new Ebola infections. In the last week of January there were 124 new cases—39 in Guinea, 5 in Liberia, and 80 in Sierra Leone—that is 25 more than in the previous week. We now have 22,894 suspected, probable, and confirmed Ebola cases and 9,177 deaths.

I am waiting for the final results with regard to T-705 or Favipiravir, an experimental Ebola medicine* of the Japanese company Toyama Chemical that French and Guinean researchers have been testing in the south of Guinea since December. Current, unconfirmed results indicate that the mortality of patients with a low to moderate quantity of EBOV in their blood decreases, but that it has no effect on people with large quantities of the virus. The phase 1 clinical trial of GlaxoSmithKline's vaccine shows that it is safe and stimulates a reasonably strong immune response. The phase 1 trials of the other experimental vaccines are still underway.

Thursday, February 12, 2015
The World Bank donates 15 million dollars so that farmers in Ebola-stricken countries can buy and plant 10,000 tons of maize and rice seed before the growing season starts.

Friday, February 13, 2015

On this inauspicious day researchers at the University of Oxford present a list of the twelve most probable threats to humanity. Ebola is on the list. The scientists do not wish to frighten people but to spur them toward more lasting development, because this is how we can prevent some disasters.

Climate change heads the list. The second threat is nuclear war. The Cold War is indeed over, but the risk of an intentional or accidental nuclear attack remains. The only way to solve this is better relations between the countries that possess nuclear weapons.

Third is a pandemic.* During the last 35 years we have had HIV, SARS, and Ebola, among others, and who knows what is yet to come? Of course, we do not have everything in hand, because no matter what there will be new outbreaks, but with prevention and permanent healthcare we can spare the world much misery, declares Oxford.

The other nine threats are a collision with an asteroid, the eruption of a supervolcano, an environmental disaster, chaos and violence due to a collapse of the political-economic system, bioterror against the human race, new weapons based on nanotechnology, artificial intelligence that escapes human control (although clever computers could significantly limit the effects of certain other catastrophes!), bad management (my greatest fear) and threats that we do not as yet know. Enough money for scientific research seems to be the best prevention of the last-named threat, and my colleagues at Oxford think so, too.

At the Belgian Embassy in Washington, air transport company Brussels Airlines advertises that: "Africa is not Ebola." With this Brussels Airlines is trying to boost the image of Africa and to tempt more passengers to fly to the continent, because the Ebola outbreak has severely reduced traffic. Although there is no Ebola in the vast majority of African countries, tourists and investors largely stay away. Bernard Gustin, CEO of Brussels Airlines, shares: "Kenya lies as far from Liberia as Brussels from New York. To stay away out of angst for Ebola is crazy." While other European air companies canceled their flights to the Ebola-afflicted countries, Brussels Airlines steadfastly remained as an air bridge for medical supplies and aid workers. But among consumers this courage earned the airline the reputation of being an Ebola-spreader.

Monday, February 16, 2015

The government of Sierra Leone promises a thorough investigation into the disappearance of roughly one third of the money that it received for the battle against Ebola.

Sunday, February 22, 2015

The Belgian B-FAST team returns from Guinea. The team's laboratory remains on site, working. The aid team began its work on December 20.

Friday, February 27, 2015

Dina and I leave for a two week trip to Botswana, but I have a holiday task: to write the last chapter of my book. That must be its synthesis, its take home message. Although I am feeling fairly fit and still have a great zest for life, I'll regard this book to be my post-scientific and social testament, so it will be good to have it properly sorted. And I now know that during the coming fourteen days I will still be following the Ebola reports. Safari or not, internet or not.

Sunday, March 1, 2015

The vice president of Sierra Leone willingly goes into quarantine now that one of his bodyguards has died from Ebola, and he calls on everyone who has met him to do the same. In Sierra Leone the outbreak seems to be increasing again. During the last weeks there was a decrease in the number of new infections, but among other places a new wave of Ebola cases has begun in the capital city, Freetown. Moreover, the quarantine of the vice president comes at an inconvenient time during which he must take on board some of the tasks of president Koroma. The president is traveling to Brussels for an EU summit on Ebola.

Monday, March 2, 2015

In Liberia the schools reopen after seven months.

Tuesday, March 3, 2015

The international community must not slacken in the fight against the epidemic, say the presidents of Sierra Leone, Liberia, and Guinea at a press conference during the international Ebola conference in Brussels. The epidemic, which surfaced about a year ago, has in the meantime claimed the lives of 9,700 people and infected 23,900. In many places the number of new infections is diminishing, but international support is also needed after the epidemic for the rebuilding of the stricken countries and in particular their health and educational services.

Wednesday, March 4, 2015
In Guinea and Sierra Leone 132 new cases of Ebola were registered during the last week of February. That is more than in the preceding weeks. According to the WHO the population in these countries does not fully understand how important quarantine and treatment are.

In Liberia, for the first time in ten months no new cases have arisen.

Tuesday, March 10, 2015
Corporal Anna Cross, a 25-year-old British army nurse from Cambridge, is infected in Sierra Leone. Cross is the third British subject to contract the disease. She will be the first Ebola patient in the world to receive the experimental treatment MIL 77 (comparable with ZMapp), and is discharged from the Royal Free Hospital in London on March 27.

Saturday, March 14, 2015
Back in Belgium. My last chapter is as good as finished.

Friday, March 20, 2015
A year after the epidemic entered the spotlight, the International Red Cross launches a campaign to make it clear that the battle has not yet been won. Due to the traffic and the close links between Sierra Leone, Guinea, and Liberia, the whole region remains a risk area so long as there is just one Ebola case. Meanwhile the number of new cases once again increases. In the West the outbreak has faded from the news. The indifference worries the Red Cross greatly. This concern gives rise to the campaign "Words against Ebola."

The current account: roughly 24,000 cases and 10,000 deaths.

I had rejoiced in the partial but spectacular solar eclipse that the cosmos ordains for this day. Alas, the sky above Kontich is blanketed by cloud. At 10:45, the highpoint, it becomes significantly grayer and darker than it already was and most of the birds in my garden fall silent. Now all the little children stand there, on all the playgrounds of the primary schools, with their little eclipse sunglasses... But then the largely covered sun peeks out between the clouds. Dina can even take beautiful photos of it.

I see the light. Ebola 2013–2015 deserves a monument, and for two reasons. Firstly: never before in 39 years of battling this virus has the cooperation between governments and businesses been so intense, and the will to develop antiviral agents and vaccines so strong. In the future this will allow us to combat Ebola better and faster. Secondly: the epidemic in West Africa has shaken the world awake and shown that an efficient world warning system is required to bring to light subsequent epidemics, of Ebola or other pathogens, more quickly. This system must be part of an international organization that has yet to be set up, which stands firmly on its own two feet financially, and which can make decisions very rapidly. It must therefore be an organization that differs in every way from the current WHO. The strengthening of basic health care systems, the training of medical school personnel, and education for every child in societies with a low or moderate income must become top priorities within the seventeen development goals that appear on the international agenda for the period 2015-2030. It sounds somewhat cynical to build a monument for an epidemic that has claimed ten thousand victims, but if this outbreak puts into motion what I hope it will, then many times that death toll will be spared in the future.

Saturday, March 21, 2015

A third team of Belgian aid workers is back from Guinea, where they tested blood samples for Ebola using the B-Life mobile laboratory. The team has ten members. In the area where they worked no new cases have been announced for more than fifty days. This third team could stay longer on site thanks to extra supplies from the Belgian Development Cooperation and Europe. "Repeatedly when people entered, blood, saliva or urine samples would be taken. In three hours we could deliver a result. In the case of a positive test treatment could be started very quickly. The result of that was that the person's chance of survival increased visibly," says a spokesman. Thus lives are saved. Soon it will be clear how many. Experience has also been gained that would position Belgium in good stead should it ever fall victim to a biological attack.

Monday, March 23, 2015

Doctors Without Borders publishes the report "Pushed to the Limit and Beyond" about the attack on the epidemic by DWB, but also by others. DWB says that the world was caught short, with the result of thousands of extra deaths. The aid organization rang the alarm bells very early, on March 31, 2014, but both the countries of the stricken region and the international community at first chose to bury their heads in the sand. In 2014 DWB committed five thousand members of staff and cared for almost as many Ebola patients. DWB could do no more, and this was extremely painful for the staff in the field. It led to great tension within the organization.

DWB urges that in the future there is a worldwide strategy for the development of vaccines, therapies, and tests. DWB closes with a balance sheet of the current situation. The epidemic is not over. In Guinea the number of patients is climbing once again. In Sierra Leone there appear to be many infected patients who do not appear on any list of dangerous contacts. In Liberia the epidemic seems to be under control, but fear of a flare-up remains great due to the situation in the neighboring countries. In West Africa there are still more new cases weekly than in any previous outbreak.

I read that 500 health workers have died from Ebola, including 14 staff members of DWB. The report is dedicated to them. Between the lines I also discern the powerlessness that the people of DWB have felt during their battle over the past fifteen months. At the end of June 2014, the organization determined that Ebola was circulating in more than sixty different places. It is impossible to take action simultaneously everywhere. And then to work in eight-part protective suits in which the temperature rises to 46°C (114.8°F). That makes enormous physical and morale demands on aid workers. Just to be able to make it through, you must do everything slowly and even breathe slowly. And if, after an hour, it becomes really just too much to bear, you must remove that suit extremely warily, or you will be infected. I cite Dr. Javid Abdelmoneim, DWB doctor in Sierra Leone (from the DWB report "Pushed to the Limit and Beyond"), during the period of September–October 2014: "We are all scared of Ebola. There is something about the way that the virus has spread itself. We try to focus ourselves on the patient, how he feels, full of misery and sickness. He feels pain and is frightened. As a doctor in your 'spacesuit' you can make little or no eye contact, but the patient also feels your anxiety. The visits that we make are also usually very short. I have limited means to help my patient. In the best case there is a bed, three meals per day, hydrating liquids, malaria tablets, and some painkillers. We especially try to let the immune system of the sick function as well as possible so that he or she can fight the virus. But as a doctor you feel yourself to be isolated from the patient. There is also enormous frustration, because the only thing that you can say is that the chance of survival is fifty percent and that beyond this there is not much you can do."

I think back on 1976 and see before me how we were clothed as we drew blood from Ebola patients in the field or were busy in the lab with Ebola-positive specimens: a laboratory apron, hand gloves, and a little mouth mask. We were the naturists of virology. And yet most of us remained uninfected. I think about courageous Sister Mariette who was spattered with vomit and blood while caring for her fatally ill colleague. Immediately rinse, disinfect, and very frequently wash your hands. And, as Father Lootens had made her swear, never touch your eyes, nose, or lips with your hands. Just try this, if you are crying, have an itch, or your nose is running.

At some moment, out of sympathy for an Ebola-infected colleague, Father Lootens wiped a tear from her eye with his bare fingers and thereafter, filled with emotion, rubbed one of his own eyes with the same finger. If this was truly the cause, I do not know, but he was lethally infected, and wished above all else that Mariette would not make the same mistake. For Mariette his advice worked.

Tuesday, March 24, 2015
The Harvard Global Health Institute and the London School of Hygiene and Tropical Medicine set up an independent panel under the chairmanship of Peter Piot, to review the global fight against Ebola.

Wednesday, March 25–Friday, March 27, 2015
During a filovirus congress in Washington DC the case fatality rate of Ebola in Guinea, Liberia, and Sierra Leone is estimated to be 70 percent. This is much higher than the most recently published WHO figures. The WHO reports 10,000 deaths from 24,000 infections, or a CFR of 42 percent. If the congress figures are right, they indicate a serious underre-

porting of the number of Ebola infections and deaths. This only strengthens my plea for a quicker witted and faster reacting version of the WHO.

Friday, March 27, 2015
Until Sunday the 2.5 million residents of Freetown and the north of Sierra Leone may not venture out onto the streets. The government goes from door to door to detect new cases and to warn people about the dangers of ritual burials. In September 2014 there was also such a campaign. In this country roughly 3,700 people have now died due to the epidemic. With approximately 12,000 cases, Sierra Leone has the greatest number of known Ebola cases in West Africa.

Sunday, March 29, 2015
Guinea has declared a health emergency for 45 days in those parts of the country that are the hardest hit. The virus has moved to the thickly populated coastal region. In Conakry, too, there are extra measures. Private hospitals where Ebola is confirmed must close. Burials may only take place under the supervision of the Red Cross or official aid services. Whoever hides sick people or transports bodies is prosecuted.

Friday, April 3, 2015
The Belgian cabinet extends the term of Dr. Erika Vlieghe as national Ebola coordinator by six months. Minister of Public Health Maggie De Block says that Dr. Vlieghe and her staff will draw up a general plan for attacking infectious diseases and will conduct research into the establishment of a specialized high security unit for people stricken by such a disease.

Tuesday, April 7, 2015
These are the last figures, and as usual they are sums of the confirmed, probable, and suspected cases. They date from April 5, 2015.

- Liberia: 9,862 cases; 4,408 deaths
- Sierra Leone: 12,139 cases; 3,832 deaths
- Guinea: 3,515 cases; 2,335 deaths

Further, there are separate figures for Nigeria (20 cases, 8 deaths), Mali (8 cases, 6 deaths), and Senegal (1 case, 0 deaths). Also in the US, the United Kingdom, and Spain there was a handful of Ebola cases.

Just before the press closed for the copy of this book on April 7, 2015, according to the most recent WHO figures there were 82 new cases of Ebola in the week of March 23–29. That is a small increase compared with the previous week, when there were 79 cases. In Liberia there were no more new cases. In Sierra Leone during the last week of March there were 25 new cases, or a decrease for the fourth consecutive weak. In Guinea the incidence increased from 45 new cases in the penultimate week of March 2015 to 57 cases in the last week.

In total there have been 25,551 cases to date and 10,588 deaths. The numbers are what they are: estimates, or—according to many experts, including those of the WHO and the CDC—underestimates. The CFR is uncertain: 41 percent overall according to the figures above, approximately 58 percent in patients admitted to a hospital. But according to an Ebola specialist at the latest Ebola congress in Washington DC the overall CFR could amount to 70 percent.

Almost forty years after the first outbreak in Sudan and Congo this is the 26th time that Ebola has surfaced, the first time in West Africa and also the first time that the spread of the virus has taken on truly catastrophic proportions. Before West Africa the situation was usually under control within a few weeks, at most a few months. By the middle of 2016 the current epidemic had been raging for 32 months. A total of 28,616 confirmed, probable, and suspected cases had been reported in Guinea, Liberia, and Sierra Leone, with 11,310 deaths. Forty-two days (two incubation periods) must elapse without new cases before the outbreak can be declared over. These periods were due to end on May 31, 2016 in Guinea, on June 9, 2016 in Liberia and on June 30, 2016 in Sierra Leone. The risk of additional outbreaks originating from exposure to the infected body fluids of survivors remains, demanding sustained counteraction by counselling on safe sex practices and testing of body fluids.

CHAPTER 7

THE LITTLE BOX

How I distill the past fifteen months during
a trip to Botswana and, in the last chapter of
my book, capture the condensed steam
of events.

Friday, February 27, 2015

And so I have finally cleared my desk, having discussed an article about the millennium goals and mentioned our flight tickets to Botswana. In the past months I have relived my life in writing against the backdrop of a huge Ebola epidemic, a scenario I would not have believed if someone had predicted it. Just as I could never have foreseen at the start of 2014, when Dina and I booked our eighteenth safari, that one year later I would be penning the last words of the manuscript of a book about Ebola, about my life as a virus hunter, about myself.

As a scientist I have always given my soul to my work, one hundred percent. I have always found it difficult to expose this soul to the outside world. That I have now bared a little part of myself is the fault of my mistress with whom I have grown old, like it or not. She bears the ten thousand deaths of this epidemic on her conscience, on top of another couple of thousand in the previous outbreaks. Behind the statistics there are fathers, mothers, children—lives that were destroyed in a couple of weeks and that ended without consolation in the obscurity of a hut in the countryside, an anonymous shack, or a shabby flat in a West African town—or in the best case in a hospital or Ebola center, with a rubber glove on your burning skin as your last contact, and as your last sight the powerless eyes, behind plexiglass, of a spaceman in a plastic suit.

Dina shakes me from my thoughts. The taxi that will take us to the airport is at the door. Since Dina and I became pensioners we have gone on safari every year in Southern Africa. I have been sold on it since 1976, and over the past decade Dina has also fallen under the spell of the continent.

On the way to Maun (north-west Botswana) I write:

The epidemic began in December 2003 in Guinea and from there it infiltrated the neighboring countries of Liberia and Sierra Leone. There was also panic in Nigeria and Mali, but with eight and six deaths respectively, damage there was limited. The few cases imported into the US, United Kingdom, and Spain, plus a couple of secondary infections, terrified these lands but did no more than that. As regards Belgium, I no longer hear them: the prophets of doom who in late 2014 forecast that it was not a question of if, but of when the epidemic would strike in Belgium. In extremis they could yet be right, but even then the cases in the western world would not be comparable in any significant way with the absolute disaster that engulfed West Africa.

In the meantime the crisis has been on the go for fifteen months, and in Guinea, at least, the tally is still rising. Why did this outbreak become an epidemic? Bitter poverty, ignorance, corruption, bad basic health care, unhygienic funeral rites, beliefs in traditional healers—all of these played together, certainly, but these factors were also present in earlier outbreaks, in Central Africa. And those outbreaks repeatedly remained limited. Also the fact that in the current epidemic the disease rapidly reached large towns was not truly unique. Granted, the former outbreaks chiefly took place in rural, remote areas, but in 1995 Ebola set up house in and around Kikwit, a Zairean town that had more than half a million residents! The balance sheet then read 315 cases and 254 deaths, a substantial outbreak, but not a large scale epidemic. Moreover, the West African epidemic of 2013–2015 did not begin in a big town, but in Meliandou, a mere hamlet in the Guinean countryside. In a later phase of the epidemic, population density would be a determining factor, but by then the seed of this sad premier had already been sown.

However, three things clearly differ with respect to previous outbreaks. First, Ebola was new in West Africa (excluding one case in Ivory Coast in the 90s), for both the patients and many health workers. Through this the virus remained completely below the radar for three months and there were tens, perhaps even hundreds, of cases by the time that the Guinean government got wind of the outbreak in March 2014. The disease had already jumped the borders into Sierra Leone and Liberia. In the Congolese Equatorial Province, where another outbreak began in August, health care was at least as rotten as in Guinea, but many health workers knew from experience that you must beware when handling a case of hemorrhagic fever, that you must draw blood samples, and where possible have these tested. Such confidence and knowledge were entirely absent during the first months in West Africa. Even after March 2014 weeks passed before health workers, laboratories, and governments were properly briefed, and months passed before the population knew the ropes just a little. By that stage there were so many sick people that the scanty health infrastructure could no longer cope, not even with the helping hands of Doctors Without Borders, who then also rapidly made it known that the situation was getting out of hand. On top of this the World Health Organization once again reacted too weakly.

The second point—I am no specialist on the matter, but this is confirmed by the WHO—is the exceptionally high mobility of people in the region. The borders between Guinea, Liberia, and Sierra Leone are extraordinarily porous. In January 2015, the WHO referred to "recent studies in which it is estimated that the mobility of the population in these lands is seven times higher than elsewhere in the world," because people travel often and far to work, to visit relatives in a neighboring country, or because the living conditions elsewhere are temporarily less bad than those in their own region.

The third explanation that I see has to do with communication. False messages are spread on a greater scale than we realize. At present the people in West Africa, despite all their poverty, enjoy much greater access to the Internet and the media in general than the victims of previous outbreaks, which took place in rural areas of Central Africa, and later Kikwit (but that was in 1995, when there was still no mention of the Internet in Africa). In Africa computers are mostly too expensive for people, but cell phones and even smartphones are often affordable. This is more and more the case in almost all countries south of the Sahara and especially in the towns where there is stable reception. According to figures released by the American opinion and data business Gallup, in 2013 at least 51 percent of the families in Liberia had at least one cell phone. In Sierra Leone that figure is 53 percent, and in Guinea 54 percent. Certainly in the towns—and in this context mass urbanization does indeed play a crucial role in the current epidemic—many people have steadily greater access to the Internet and television. This should be a great advantage for health communication, unless false messages are being spread, of course.

Because the disease slumbered for so long, at first it didn't seem to penetrate people's consciousness and it was often obstinately denied—one reason for this may be that the messages were as hard as nails and hopeless. They boiled down to the following: "There is a new disease called Ebola and it is extremely serious and deadly. No vaccine exists for it and there is no treatment, thus do not touch each other and go to the hospital or a medical aid post if you are sick." The clear message was intended to rub people's noses in the facts so that they would adjust their behavior: you report to the hospital or a medical center if you have a fever, inform the authorities of suspect people, keep your distance from others, bury the deceased safely. But in many cases it had precisely the opposite effect. People fled in panic or lost their courage. "I will definitely go to my traditional healer. Why should I hand myself over to western medicine as it says itself that there is no treatment? If I am almost certain to die, then I would prefer to do so as comfortably as possible, at home, among the people who love me and the faces I know. Then I am not leaving myself in one way or another to the oafs in yellow plastic suits. And what if I have a fever and it is due to something else? I am not going to spend any days in a transit center that I enter without Ebola and perhaps leave with Ebola, probably in a body bag. And if I take my mother, father, or feverish child to such a center, then there is no one there who will let me know how it is going with him or her. Then I won't even know myself if my loved one dies and is buried somewhere." Here I conjecture as to what I believe people think, for what it's worth. In any case, I wonder if it is still better not to communicate, than to communicate out of despair only to strengthen the stigma of the disease.

From day one a cleverer message would have been: "There is a new disease called Ebola that you must always take seriously, but which half of all sick people can survive provided they allow themselves to be cared for in a hospital or treatment center. There you have a real chance. Outside, the chance of making it is very small. It is better not to stay at home if you are sick, because you can infect other people there." People will only believe that last sentence once whole families have been wiped out by Ebola at their homes. Of course, it makes sense that in many places there were too few hospital beds for weeks and that patients had no option other than to stay at home, thus better communication would have helped little. But even after international aid had fully powered up and there were enough beds, in some areas patients remained hidden at home or elsewhere.

Nigeria tackled it better right from the start. When I congratulated my friend Dr. Oyewali Tomori in October concerning the intervention of the Nigerian government, above all else I wished to give him a fillip. From an article that Tomori and his colleagues had published in *Eurosurveillance* about the Ebola outbreak in Nigeria in the period July–September 2014, I inferred that the Nigerians had succeeded completely in tracking down the clinically suspected people, in testing possible cases with a sensitive and specific test to confirm or negate them, and in finding, isolating and monitoring people who had been in contact with patients as quickly as possible. Hence Nigeria could limit the number of sick people, and not only support them but also really treat them therapeutically. Moreover, the medical personnel were properly trained and protected by the right clothing. The mortal remains of the people who had died of Ebola were interred with dignity, but above all else with safety.

All of this is not so difficult, but if the disease can run riot for three months, then it has spread so much that the point of no return has already been reached when you are still just starting to track cases. Moreover, local residents have already seen so many deaths around them that panic breaks out and people hide away or flee in all directions, whereby the epidemic expands yet farther, setting up a negative spiral. Then the only thing that can still help is massive aid from abroad, and a sustained fight of months or years, certainly if you don't have any widely available means of treating the disease and vaccines.

Of course prevention is best, but in this regard you can only hope for better basic health care and enlightenment, a preventive vaccine, or... that you find the host of the virus! Research has shown that EBOV can survive in three species of fruit-eating bats without killing them: *Hypsignatus monstrosus* (the hammerhead bat), *Epomops franqueti* (Franquet's bat), and *Myonycteris torquata* (the collared bat). The virus has yet to be found living in these species in the wild, but genetic material (RNA) has indeed been found. The three species are present in all the African lands where Ebola has ever been identified: The Democratic Republic of the Congo, Guinea, Liberia, Sierra Leone, Uganda, Ivory Coast, Gabon, Congo-Brazzaville, and South Sudan. Via satellites we can delimit the habitats of these species and thus identify Ebola risk areas. Hence, in addition to the countries that I have mentioned, you must also think about parts of Nigeria, Cameroon, the Central African Republic, Angola, Togo, Tanzania, Ethiopia, Mozambique, Burundi, Equatorial Guinea, Madagascar, and Malawi.

During the last 39 years the population in the countries at risk has nearly tripled, from 230 to 639 million. On top of this, the fraction of people who live in towns (and not in the countryside) has risen from 25 to almost 60 percent. Add to this increased mobility and you may be sure that a new Ebola outbreak or epidemic will happen, one that will perhaps spread even more quickly and farther than the current one. Worldwide there are approximately 1,200 species of bats, of which only 15 percent have been investigated for viruses. Precautions are thus the order of the day. We have also found EBOV in gorillas, chimpanzees, and duiker antelope, but these animals are not filovirus reservoirs, because they become sick or die of the infection.

Regardless of all else, we will never be able to properly eradicate Ebola, as we have eliminated smallpox, unless we find the reservoir and wipe out all its members—whereby we would disrupt ecosystems and perhaps expose ourselves to zoönoses that are even worse than Ebola. But if we were certain of the reservoir, we could, of course, inform people and thus prevent outbreaks and save lives. Sometimes when I again reflect upon the fact that I never succeeded in finding Ebola's animal host, it gnaws at me. But even with the most modern methods my successors have been unable to isolate the virus in even one of the numerous bat species that must be considered a possible reservoir.

Can we learn something from this epidemic? One of the most important lessons is that you must involve the population better in the fight against Ebola. That can be done by organizing information campaigns that make use of traditional (radio) as well as modern means of communication (cell phones and the Internet), and that take into account cultural sensitivities such as superstitions and myths, and which do not so much make the people afraid or stigmatize the sick, but motivate them to adapt their behavior and reward them when they do so. Political authorities, religious leaders, village chiefs, local herbal healers, and convalescent Ebola patients must also do their part. People who have beaten Ebola can make a real difference. They can tell their fellows a success story, can convince them that the "foreign, strange help" did, in fact, make a difference. Some sick people are transported to a treatment center in a state of high anxiety, even against their will, but through adequate treatment they return home, healthy. They have also seen that other people who were sick but tested negative for Ebola could once again breathe easily and obtain medicines for what they actually had. A fuss must be made of cured patients when they leave treatment centers; they must receive a certificate on which it is stated that they are Ebola-free, that they pose no danger to family and friends, and that they know better than anyone else what must happen if another loved one should fall ill.

Monday, March 2, 2015

I wake up in our lodge and I look out at a magnificent baobab tree. Yesterday I strolled around this miracle of nature and estimated the circumference of the giant to be roughly ten meters. According to the people here, that corresponds with a lifespan of 2,500 years. We will never know for certain because baobabs do not have annual growth rings. During the rainy season the tree sucks up so much water into its trunk that it can easily survive the following drought.

This evening I install myself beneath this stout comrade to attack my writing. Many governments could learn a lesson from far-sighted trees. If and when we succeed in bringing the epidemic completely under control, the afflicted states must have a plan and a lot of money to come out on top. It is now or never if they wish to set up an expensive health care system, give a thorough training to aid and health workers, and restore social life. Ebola will function as a pilot flame that must encourage managers to invest in health and education. Sooner or later—this we can assume—hell will break loose again, and hopefully Guinea, Liberia, Sierra Leone, and all the other lands in the risk zone will be in a better state to deal with it and limit the damage.

Meanwhile history has been made. Ebola can—forever more—never again be regarded as a disease that only strikes relatively small groups of people in remote Central African hamlets. Whatever figures you believe, the real numbers of infected and dead people are probably greater than those that appear on the websites of the WHO and the CDC, if only because the end is still not clearly in sight.

Many hundreds of health workers have died, and then we haven't even considered the countless "semi-official" carers, relatives, and friends who have departed this life because they nursed a loved one.

Alongside the direct death toll there is collateral damage, firstly to the health care system itself, because Ebola gobbled up people and resources that could not then be given to other medical problems. People also died because of this problem, or became sicker—a situation that would never have happened without Ebola. Secondly, there is the collateral damage to agriculture and the economy, by quarantines, unsuitable labor, the implosion of trade, and disrupted distribution networks. Thirdly, there is the impact on interpersonal relationships of a disturbed social life, stigmatization, and institutionalized suspicion. Ebola prompted another thing that has never happened before: the small but constant evacuation of EBOV infected patients—especially repatriated health workers—who receive the best possible medical treatment in rich countries. Many people find this dishonorable or say that the media attention lavished upon these cases is excessive, but we can learn an enormous amount from them, about the clinical course of the disease and the possible efficacy of specific therapies.

The top priority for science, health organizations, and the pharmaceutical sector is the improvement of techniques to detect Ebola and to treat infected people. The WHO has begun to evaluate new and rapid diagnostic tests that are user-friendly, sensitive, and specific, and which can be performed in the field, if necessary, in bad circumstances and remote areas. Hence, people could be tested and treated in the same place (point of care testing), at least if the investigation of antiviral agents is pursued, just like the testing of vaccines.

Friday, March 6, 2015

During the previous days we flew from Maun to Kasane, the northernmost place in Botswana, where the four countries of Botswana, Namibia, Zambia, and Zimbabwe meet at a point. We have also chalked up a boat trip on the Chobe, which took us through poetic landscapes with elephants playing close by and the most wonderful waterbirds. At the safari camp we have become acquainted with Morgan, our guide for the coming days. Morgan is a walking nature book and this is not at all strange, because his grandmother was the local and universally valued herbal healer. He tells us how she packed a large space next to her thatched hut with all sorts of concoctions that she prepared day and night. In this veldt pharmacy hung a mass of dried herbs and plants. On iron gratings above a smoldering wood fire stood simmering copper kettles. She even had a bamboo distillation column to which a sloping condensation tube was secured by means of old bicycle tires. Morgan recalls how as a young lad he watched, fascinated, as the crystal clear liquid dripped out of the long pipe into an empty gin bottle.

He was also fascinated by the little wooden box that hung around the neck of his grandmother. It was made of mahogany, which shone due to years of being gripped and rubbed. Grandma used this little box when sick people asked her for help. She would lay her left hand on the forehead of the patient and grasp the little box in her right hand. Subsequently she spoke incomprehensible sentences to make contact with the ancestors, which in turn gave advice. Regularly she lifted the little box to one ear and then shook it to and fro. Sometimes it made a rattling noise, Morgan had noted, but not always. Yet the little box could not be opened.

When his grandmother died, Morgan hung the trinket around his own neck. He still wears it, under his shirt. He is a man of nature, with his feet planted firmly on the ground, but his grandmother's little box is a magical object, which Morgan even now suspects has saved his life on certain occasions, or at least favorably influenced events. Morgan regularly tries to make contact with his grandmother. To do so, he holds the little box to one ear and shakes it. Now and then he hears something rattle.

As a western scientist or health worker you can shrug your shoulders at so much superstition, but that will take no wind out of the sails of people such as Morgan's grandma. Whatever you may try, there is ongoing respect for herbal healers in the hope that with time this respect will be repaid, and that at the very least you will not have such an influential person working against you.

I am happy that I can also tell Morgan a tale about a little box, even if it is perhaps not as magical as that of his grandmother's. In 1994 I got to know Rudi Pauwels, a young pharmacist who had completed a doctorate a few years earlier on new agents against HIV at the Rega Institute for Medical Research in Leuven. At that time Rudi had just set up Tibotec, together with his wife, Carine Claeys. This is a pharmaceutical company which at first focused on research and development of treatments for HIV, and later also for hepatitis C. A year later they and Paul Stoffels would create Virco, an HIV diagnostic company that would develop technologies to help select the optimal anti-HIV drug regimens. In 2002 the combined Tibotec-Virco group was acquired by the American pharmaceutical giant Johnson & Johnson. But after heading the anti-infectives group for 2 years, the then 43-years-old Rudi felt he needed to pursue another dream. Thus he departed for the École Polytechnique Fédérale in Lausanne, Switzerland, to discover new technologies there that could reduce an entire laboratory to... a little box, not bigger than a dish sponge! The dream never left him, and in 2007 it led to the establishment of the company Biocartis, and his dream has now become a reality.

While I tell Morgan my tale, it has become dark. He switches on an electric lamp made in China, which is powered by solar energy. For this you do not need a noisy generator. He produces an ice bucket with a bottle of South African white wine. "Klein Constantia" appears on the label. Dina, whose official name is Constantia, tastes the wine twice and opens her laptop. In just a few moments this young woman of 72 has found a photo of the "Idylla" on the internet, the machine into which the magical little box of Rudi Pauwels is inserted, about the size of a laser printer. I turn the photo toward Morgan and tell him that I could put this equipment on the camp table, and with a single finger prick put a drop of his blood in the magical little box, that I would then insert into a small opening in the Idylla. This apparatus would then search for 30 different viruses, bacteria, parasites, and biomarkers at the same time, and in exactly 90 minutes I would be able to read the results on the machine's screen. If I wish, the data will be sent wirelessly, via satellite, around the world to centers where they will be analyzed for epidemiological research. It looks too good to be true, hence the name "Idylla," but the machine exists and it is available.

Wednesday, March 11, 2015

I will invite Rudi Pauwels to the presentation of my book and ask him to make another little box. One that I can use to carry back to cold Belgium not only the awe-inspiring landscape and the earsplitting croaking of the frogs on this trip, but also the indefinable evening glow, the soft wind on my skin, the smells of the sand and of the flowers, and the slow pace of life that requires no explanation. I get up to fetch the single article that I have brought with me to Botswana, about the millennium goals.

The Ebola epidemic in West Africa has otherwise demonstrated how great the need is for leaders who can act energetically and develop a long-term vision for everyone's benefit. Since 1990 the United Nations has been debating the growing gulf between the North and the South with respect to development and welfare. In September 2000, 191 government leaders sat round the table in order to set eight goals that should be realized during the beginning of the new millennium, by 2015: goals connected with poverty, education, child and mother mortality, the combatting of deadly diseases, safe drinking water, and development aid.

We are now in the year 2015 and the goals that were set—this won't surprise you—have not been achieved. Yes, progress has indeed been recorded and I see the glass as being half full rather than half empty. Through international cooperation, 700 million people have been freed from extreme poverty. The world was able to prevent 3.3 million people dying from malaria, and even more lives have been saved through the successful treatment of tuberculosis. Since 1995 access to antiretroviral therapy has extended the lives of 6.6 million HIV-infected people. Also real, but less easily measurable, are the strides taken toward better access to basic health care and a cleaner environment. In the North, the awareness of people with regard to the environment has increased spectacularly in every respect; laws have been passed to protect the environment and businesses have invested in sustainability and innovations to reduce their ecological footprints.

Socio-economically the funding gap remains, and unemployment in the South is dramatically large, but there is evolution. A middle class is arising in ever more southern lands where earlier only a mass of desperately poor people and a minimal percentage of superrich lived. The region of Lagos in Nigeria is an example of a densely populated African area where the middle class is a steadily growing reality. Africa will soon be an enormous consumer market. Eleven African countries can be found in the top 20 fastest growing economies, but among the top 20 poorest countries Africa is represented eighteen times! Five African states—Uganda, Congo, Ethiopia, Mozambique, and Sierra Leone—even appear in both lists. In other words: if you are coming from nowhere, then any growth is spectacular.

During the last years the rich North has also had its problems, with a crisis in the non-perishable goods sector and the banks. Also within the European Union a north-south gap has made itself felt, which has yet to be resolved. Perhaps our problems do not appear to be significant compared with those of the three Ebola-stricken countries, but if rich countries must economize, of course that never plays out in favor of development aid or solidarity with the South.

In September 2013, an open working group of about thirty leaders was formed to formulate the so-called Sustainable Development Goals (SDGs), with the agreement to realize these by 2030. Never before in world history had such broad international consultation occurred about subjects that are critical for a peaceful and healthy planet. One of the SDG's concerns is a guaranteed healthy life with well-being for everyone. That implies that infectious diseases such as AIDS, malaria, and tuberculosis must be combatted, but meanwhile the non-transmissible diseases* (NTDs) have become an equal priority. Of the 53 million people who died in 2014, 65% succumbed to an NTD: heart disease (48 percent), cancer(21 percent), chronic lung disease (12 percent), diabetes (3.5 percent), and obesity. I would also add dementia to the priority list, a disease that has a devastating impact on the patient and the family. Worldwide the number of dementia cases is estimated as 36.5 million, and that number will double by 2030.

Also in the South, "diseases of wealth," such as narrowing of the coronary arteries, obesity, and drug addiction, are of increasing importance. Not because the people have now garnered so much wealth, but because they have swapped the Spartan subsistence of the countryside for a life in the big town, where you do not need to cripple yourself with work in order to pull a few tubers from the ground, where you do not have to walk many kilometers in order to fetch more or less clean water, where there are no snakes and crocodiles, where people wear western clothes and dance to the music of Lady Gaga, where you can prostitute yourself, where there are sometimes work opportunities and things that can be sold, where crack, crystal meth, and other drugs circulate, and ever more people eat the same fatty or sugary junk food as their counterparts in the rich North.

Our ranger here, Carlos, is a powerful black chap, but he eats like a sparrow. Just as it was once with us, to be fat in Africa is now often a status symbol. Fat means eating and eating means money. Carlos had shown off his status with so much conviction that his heart had almost failed. Now he controls his diet and doesn't drink any more cola, even if that beverage has the local reputation of heightening potency.

And then there is the resistance of ever more bacteria, viruses, parasites, and fungi against drugs. An intelligent, globally coordinated, and comprehensive action plan is essential to detect resistance and to take measures. Otherwise the list of pathogens that we do not have a single medical defense against will grow longer and longer. Again, fast and accurate diagnostic tests are essential. This year and in the forthcoming years it would be very useful if modern, mobile, and durable equipment is available in the countries where Ebola has appeared and where potential bat reservoirs of the virus are present, so that rapid tests can be performed.

Apart from Biocartis, the companies Luminex, Cepheid, and Becton Dickinson are also working on this kind of solution. It seems logical to me that in the first instance one tests simultaneously for the presence of EBOV (Ebola-Zaïre), SUDV (Ebola-Sudan), *Plasmodium falciparum* (malaria), and Lassa. Ebola, malaria, and Lassa are present in the same areas and share a number of symptoms (fever, nausea, headache, fatigue). Due to this they are often confused with each other. My successor at the ITM, Kevin Ariën, even has a list of thirty different viruses that can cause fever in people who have returned from Africa.

With the technology in devices such as the Idylla, the resistance of viruses against antiviral products can also be determined, so that after 90 minutes the patient can at once receive correct and effective medication. Amid the steadily rising costs of health care this would offer a substantial saving of both time and money. The recently developed BRAF Mutation Test also gives one hope. As has been mentioned earlier, in 2014 60 percent of all deaths worldwide were the result of non-transmissible diseases. Cancer occupies second place in the list of NTDs. The new Idylla test permits extremely early detection of cancer markers. This means that long before a tumor has developed in a patient's body, researchers can find a few specific mutations in that person's genetic material that correlate with the appearance of specific cancers. In the future that will allow us to treat cancer very early. There shouldn't even be a visible tumor.

By 2018, so we hope, polio will have been eradicated with the support of the Bill and Melinda Gates Foundation. That would be truly concrete. But in the background I hope that I too, will be there in 2030 as a man in his late 80s, to be able to see if humanity has been able to realize the SDGs. Then I can really retire. In the meantime, I am turning the key on my archive cabinet and giving it to Jana, one of my granddaughters. Next week she must give a talk at school and it will be on... (I didn't whisper it in her ear) Ebola. Jana will definitely do an outstanding job.

GLOSSARY

This list contains explanations of concepts,
names and abbreviations that are used in the book,
but it can also be enjoyable just to dip
into this information here and there to pick up
some serendipitous factoids.

ABSTRACT
Summary of a scientific article.

AIDS
AIDS is an acronym that stands for acquired immunodeficiency syndrome. I will unravel this:

"Acquired" means that you get the disease during the course of your life and that the problem does not arise from within the body itself. The disease is not congenital. That is a little confusing because if a pregnant woman has the AIDS virus, the fetus can be infected with the virus through the mother or at the moment of birth. In that sense, the problem does indeed look as if it is "congenital," but it is not in the baby's genetic code. Moreover, the baby is only HIV-positive, which is different from having AIDS.

"Immunodeficiency" describes an immune system, or defense, which is "deficient," i.e. isn't working as it ought to.

A "syndrome" is a group of signs of sickness (or symptoms) that are present together and that can have multiple causes. Sometimes the cause is unknown. This was the case when the name "AIDS" was conceived. Since 1983, we have known that AIDS is caused by the human immunodeficiency virus: HIV.

Although thanks to drugs HIV no longer need be a death sentence, AIDS is nonetheless a worldwide health problem. To date at least 39 million people have died from the disease, especially in Africa. In 2013 approximately 1.5 million people died of AIDS.

WHAT DOES HIV DO?
HIV invades and destroys the T-helper cells, which regulate the immune system during an infection. The virus does this to multiply itself and at the same time neutralize the immune system of its host. This can be a highly effective offensive strategy: you invade a country, recruit new soldiers there, and in passing shoot down the border guards. Not only HIV profits from this, but also other disease-causing agents. These are given free rein, and usually it is one or more of these opportunistic diseases that eventually kills the patient.

Investigation shows that the disease that appeared on the American West Coast at the end of the 1970s had earlier signaled its presence in Africa. In Europe, too, for that matter. At the end of the 1930s there must already have been an AIDS epidemic in Poland. Now we estimate that the syndrome must have arisen around 1900, at the same time as most African towns. The disease could spread more easily through the concentrations of people. Originally the virus was present in monkeys. It probably jumped across to humans via the hunting and sale of monkey meat. A large increase in the number of people cutting wood in the forests - people who rely heavily on bushmeat - would have increased human exposure to monkeys and the virus.

IMPACT IN THE WESTERN WORLD

Regardless, partly through the death of celebrities such as Rock Hudson and Freddie Mercury, AIDS grew to be the world's second most discussed disease, after cancer. But cancer "conquers" you. AIDS, by contrast, would for a long time be regarded as a punishment earned by engaging in homosexual activity or by injecting drugs into a vein. When it became apparent that AIDS is first and foremost a tropical disease and a large-scale problem for young, poor heterosexual African men and women, it became a priority for institutions such as the Institute for Tropical Medicine in Antwerp and the United Nations. That led to the founding of the special UN program UNAIDS, which was chaired by my friend and colleague Peter Piot until the end of 2008.

TRANSMISSION

HIV has caused a pandemic and yet it is not so easily transmitted. You cannot pick up the virus by sitting in the same room as infected people or by touching them, hugging them, French kissing them, or drinking from the same glasses. You can only catch the virus if one of your mucous membranes (for example, inside your mouth, the glans of the penis, the anus, or vagina) or your circulating blood comes into contact with specific body fluids of an infected person. These fluids are blood, semen, pre-ejaculatory seminal fluid, vaginal moisture, and mother's milk. The risk of transmitting this virus through other body fluids—saliva, sweat, tears, urine, sebum, pus, ear wax, and diarrhea liquid—is impossible or negligibly small.

Usually transmission occurs during lovemaking and more particularly during intercourse, anal intercourse, rimming, and fisting. Excuse my French, but we must call these things by name. In addition you can get HIV through a transfusion of infected blood or the use of infected needles (from injecting drugs, for example). Moreover an infected mother can give the virus to her child during pregnancy, at birth, or when she breastfeeds the infant. HIV is not by any means always transmitted when there is risky contact. Yet AIDS has become a terrible health problem. This is due to all sorts of reasons.

LUST

The first factor is that we love to make love. We do it so often and sometimes so rashly that even a sexually transmitted disease with a low risk of transmission through sexual contact can cause a true epidemic. Nowadays, most people have multiple sexual partners, mostly consecutive (in serial monogamy), but in 30% of cases also overlapping. Condoms could have spared humanity much misery, but due to all kinds of cultural obstacles—and a lack of good public information or distribution channels—for a long time condoms have been underused, especially in African countries south of the Sahara. Banning condoms because they might encourage dissolute sexual behavior was not a good idea. In this regard the Catholic Church bears an overwhelming responsibility.

DEATH PENALTY
A second factor is that in the first fifteen years after the (re)discovery of AIDS, there were simply no usable therapeutic agents that could inhibit an HIV infection or force it into remission. For a long time AIDS was a disease from which you died, on average nine months after the first symptoms.

PSYCHO VERSUS MACHO
The third reason for this global disaster is the furtive slowness of HIV. Like every other virus, HIV has an incubation period, but in the case of AIDS on average and most often nine to ten years pass between infection with the virus and the appearance of the sickness. Only after the incubation period do you have AIDS, before this you are just seropositive. This is what is so treacherous about HIV: it has an ocean of time to infect other people, unnoticed. Ebola is macho, HIV is cunning.

NOW YOU SEE ME, NOW YOU DON'T
There is still no vaccine against AIDS, however hard we have searched. To date, all attempts to make a vaccine and thus to promote the production of antibodies have failed. The antibodies always work but only against one form of the protein envelope around the virus, and before you know it that envelope has changed.

TAKE HEART
It is still not possible to cure AIDS definitively, but nevertheless, there are good HIV-inhibitors. These can prevent those infected with HIV from developing AIDS, and if they already have AIDS, HIV-inhibitors can delay its progression. Current treatment consists of a combination of inhibitors and it is called Highly Active Anti-Retroviral Therapy. HAART is very effective and has become easier to sustain. Under pressure from AIDS organizations, the pharmaceutical industry decreased its prices for HIV-inhibitors, and due to this, ever more people are able to access the treatment. The fact that an HIV infection can now be treated also has a disadvantage: HIV disappears from the news, people think that it is no longer a problem, and they once again begin to indulge in risky behavior. Thus the number of new infections is again climbing in western countries.

ANTIBODIES
A substance that is naturally present in an organism, directed against an antigen that is naturally present, or a substance that is produced by an organism in reaction to an antigen that has penetrated the body of the organism.

ANTIGEN
A substance that activates the body's defense system.
    See: defense, monoclonal antibody

ANTISERUM

See: sera/serum

BDBV

Bundibugyo virus, a species of Ebola virus
    See: Ebola virus

BIOSAFETY LEVEL (BSL)

The degree of biosafety for which a laboratory has been equipped. BSL ranges from 1 to 4.
    See: maximum security laboratory

BLOOD

In higher animals—and yes, despite everything we do, that includes humans—blood is a fluid that circulates through the body to remove carbon dioxide and waste materials, to maintain body heat at its target temperature, and to distribute oxygen, nutritional substances, and defensive substances. Blood is pumped by the heart through a closed circuit of arteries, capillaries, and veins. Cells and other components are found in blood. There are three types of blood cell: RED BLOOD CELLS or erythrocytes (in which hemoglobin, especially, is found—for the transport of oxygen and carbon dioxide), WHITE BLOOD CELLS or leucocytes (for defense), and BLOOD PLATELETS or thrombocytes (so that the blood clots if you have a wound). The other components are MINERAL SALTS AND IONS (such as sodium chloride, sodium bicarbonate, potassium, magnesium, and calcium), countless TRANSPORT PROTEINS (albumin tops this list, as it draws water into the circulatory system and so regulates blood pressure), ENZYMES (such as clotting factors), CHEMICAL MESSENGERS (among which are all sorts of hormones), and ANTIBODIES. And to conclude, blood also contains NUTRITIONAL SUBSTANCES (such as glucose, cholesterol, and oxygen) and WASTE PRODUCTS (such as carbon dioxide).

A typical person has nine to thirteen pints of blood, of which 60 percent is PLASMA and forty percent cells. Plasma is the liquid component of blood, containing the many substances that are dissolved within it. Plasma—blood without the three types of cells—is pale yellow or grayish yellow. The blood cells are not dissolved in blood, they are suspended within it (they are carried around by it).

Blood is drawn for laboratory tests. An anti-clotting agent is immediately added to this blood. Thereafter it is placed in a centrifuge, which separates the plasma from the cells. If plasma is frozen it can be preserved for a virtually unlimited period for research purposes.

Blood clots outside the body. A thin, yellow layer appears on the surface of the clot—this is SERUM. Serum is blood without blood cells and without clotting factors. Both plasma and serum can be used for many laboratory tests, but serum is in principle better because the clotting factors in blood plasma can influence the test result.

BOB

The Belgian Bewakings- en Opsporingsbrigade (literally: Guarding and Detection Brigade). In the media this was often wrongly called the Bijzondere or the Belgische Opsporingsbrigade (the Special or Belgian Detection Brigade). Until the end of the 1990s this was a curious part of the Belgische Rijkswacht (the Belgian National Guard—the country's police force). "BOBers" conducted interviews and undertook detective work and surveillance. To the majority of the public the detectives were fairly grim figures with hats and long raincoats, who constantly drove around in little Renault 4's or Volkswagen Jettas, so systematically inconspicuous that through this alone they were conspicuous—although that might also have been due to an antenna like a fishing rod on the car roof. Following the police reform bill of 1998, the BOB was absorbed into the Federale Gerechtelijke Politie (Federal Judicial Police). The BOB should not be confused with Staatsveiligheid (State Security). I do not know if I really was shadowed during the Cold War due to my contacts with Russian and American virologists, and if so, whether this was by the BOB, State Security, or both services. In any case, one fine day I was hauled over the carpet, and that was at the BOB.

BSL
Biosafety level.
See: maximum security laboratory

CASE FATALITY RATE (CFR)
Or case fatality ratio: the ratio between the number of cases of a disease and the number of people who die from it (fatalities). It should actually be "fatality case rate," because the numerator refers to the number of deaths and the divisor to the number of cases. If the CFR of a disease or epidemic is 60 percent, this means that 60 percent of the people who get the disease also die from it. In the Netherlands this is also called the lethality or deadliness of a disease. The CFR should not be confused with the death figure (or mortality), because that is the annual number of fatal cases per thousand people.

CCHF
See: Crimean-Congo Hemorrhagic Fever Virus

CENTERS FOR DISEASE CONTROL AND PREVENTION (CDC)
The CDC is an American state institution with maximum security labs (MSLs) and an outstanding reputation in the area of infectious diseases. The center, which is also involved in health education, is housed in Atlanta, Georgia, and was founded in 1946 as the Communicable Diseases Center. That happened in the wake of a program to combat malaria during World War II. In 1970 it was renamed the Center for Disease Control and in 1992 received the name that it carries today. Through all this the abbreviation CDC remained in use. The European Union's equivalent of the CDC is the European Centre for Disease Prevention and Control (ECDC), which is housed in the Swedish capital, Stockholm.

CFR
See: case fatality rate

CHLORAMPHENICOL
An antibiotic against a broad spectrum of micro-organisms that is now less used, due to its side effects (among which is anemia), with the exception of some potentially seri-

ous infections such as abdominal typhus, cholera, and bacterial conjunctivitis. The early clinical symptoms of patients with Ebola resemble those of typhus, hence treatment of the new disease with chloramphenicol was sometimes started.

COMPASSIONATE USE

Sometimes a medication that has not been approved for one reason or another is nevertheless administered to one or more patients because their quality of life has greatly diminished, or because they are at significant risk of dying and there is no better therapeutic alternative. Thus drugs are occasionally given to patients despite the fact that they have undergone no testing or incomplete testing on humans.

A variant of this is MEDICAL NEED, the use of an approved medication for a different indication than that for which the medication has been approved. Compassionate use can sometimes save people and deliver data that contribute to the approval of a medication or accelerate the approval process.

See: Ebola medicines

CONGO

See: Democratic Republic of the Congo

CONVALESCENT/CONVALESCENCE

Recovery, the phase before complete cure. Convalescence is the gradual process by which someone regains their health after a disease or injury. The word can also refer specifically to the late phase of an infectious disease during which the patient is recovering or feels better, but can still be a source of infection for others. For example, Ebola can be shown to be present in the sperm of a sick man for up to six weeks during recovery. Remains of the virus were found in the sperm of Geoffrey Platt, the scientist infected at Porton Down in 1976 with SUDV, even when he was significantly better.

COTRIMOXAZOLE

An antibiotic against a great number of bacteria, among which is Salmonella typhi, the cause of abdominal typhus.

CRIMEAN-CONGO HEMORRHAGIC FEVER VIRUS (CCHF)

Transmitted by ticks which are especially abundant in Turkey. The name of the virus refers to the places where the disease was discovered: in the Crimea in 1944, and in the Belgian Congo in 1956.

CRIMEAN HEMORRHAGIC FEVER

Now called Crimean-Congo hemorrhagic fever.

See: Crimean-Congo hemorrhagic fever virus.

CULTURE MEDIUM

See: nutritional substrate

DEADLIEST DISEASES: TOP 10

Ebola, breast cancer, and colon cancer are often in the news, but by a wide margin they are not the deadliest diseases on earth. Even malaria isn't in the top 10. An explanation is in order. By "deadliest diseases" I do not mean those diseases that carry the greatest risk of death once you have contracted them, however small the chance of getting the disease may be. If that were the criterion, then Ebola (EBOV) indeed scores very highly (50 to 90 percent deadly, regardless of the outbreak), together with other killers for which no true means of treatment exist, such as the prion disease Creutzfeldt-Jakob (100 percent deadly), primary or granulomatous amoebic encephalitis (90 to 100 percent deadly), Lujo (80 percent deadly), herpes B (70 to 90 percent deadly), influenza A subtype H5N1 (60 percent deadly), and Marburg (20 to 90 percent deadly). Further, there is a series of diseases or disease-causing agents that can be extremely deadly if you encounter them, but for which an effective treatment indeed exists, although it is not available everywhere and at all times, is not always affordable, or cannot reach you quickly enough. Among others this is true of sleeping sickness, the plague, rabies, leishmaniasis, AIDS, and anthrax—all diseases that in the absence of good treatment leave little to no chance of survival. Only one of these diseases appears in the list of the deadliest diseases that appears below: AIDS.

The following top 10 consists of diseases that can be deadly if you do not handle them well, and that also strike many people, for example through bad hygiene or an unhealthy lifestyle, or due to the infectivity of the disease (as in airway infections). Aging also plays a role. Health care has drastically improved compared with just fifty years ago, and proportionately fewer people die from hunger, traffic accidents, and world wars. Therefore the average age of everyone on the planet rises, but nevertheless we all die somehow (often from cancer or heart disease). This phenomenon also influences the statistics. The result of all this is that worldwide the following diseases kill the most victims.

- At number 1 stands CORONARY ARTERIAL DISEASE. If the arteries that must supply the heart with oxygen become gummed up by smoking, a bad diet, or too little exercise, at some moment the pump receives too little fuel and it stops. Result: more than seven million deaths worldwide each year.

- At 2: STROKES (cerebrovascular accident or CVA), a blood vessel in the brain that is blocked or leaking, whereby brain tissue dies within minutes. Strokes cause about six million deaths per year.

- Tied at 3 and 4: CHRONIC LUNG DISEASE or COPD (especially from smoking) and INFECTIONS OF THE LOWER AIRWAYS (among which are bronchitis, lung inflammation, and flu). Both COPD and infections of the lower airways claim about three million lives per year.

- At 5: CANCER OF THE LUNGS AND AIRWAYS. Chief cause: smoking. Almost three million deaths per year.

- Tied at 6, 7, and 8: AIDS, DIARRHEA, and DIABETES, each of which repeatedly claims one and a half million lives per year.

- At 9: complications due to PREMATURE BIRTH OR EXCESSIVELY LOW BIRTH WEIGHT, with about one million deaths.

- At 10: TUBERCULOSIS, with almost one million deaths per year.

This list puts the disconcerting figures for the West African Ebola epidemic into perspective. Of course one can argue about any ranking. It all depends what you take into account as the ultimate cause of death, what categories you define, and from which years the figures are drawn. Moreover epidemiological details from different countries are not always comparable. Furthermore, there are also differences between more and less developed countries. So it is that AIDS appears in the world list because this disease still claims third place as a cause of death in developing countries, and people in rich countries seldom die of diarrhea. What is striking is that narrowing of the coronary arteries—often regarded to be a disease of wealth—is in first or second place overall, also in lands where people have a moderate or low income. Airway infections and strokes also score worryingly high globally. Furthermore, cancer is increasing everywhere, in particular because it is chiefly a disease of aging and everywhere in the world we are on average becoming older. In the poor countries, where a quarter to a third of all deaths strike children younger than 14, far too many people die due to other diseases well before they have the "chance" to develop conditions that are often connected with an advanced lifestyle. But even there cancer is on the increase.

The relatively up-to-date list above refers to the whole world and I have based it on an article from *Healthline* that appeared online in September 2014. I have checked the data and they agree broadly with the most recent data of the WHO. He who wishes to analyze the facts himself will find extra details and material to reflect upon at www.who.int/features/qa/18/en.

DEFENSE

The ability of an organism (such as a human) to defend itself against disease-causing agents (pathogens). Also called immunity. "To be immune" or "to have immunity" means that you are vulnerable to a specified sickness (agent) but that you are protected against it. There are different types of defense. Thus you have hereditary immunity and immunity that you gain during the course of your life. There is also specific defense (against a well-defined cause of disease) and non-specific defense. The different ways in which the body detects causes of disease and neutralizes them forms a complex system: the defense system or immune system. I will outline the system.

PHYSICAL, CHEMICAL AND MECHANICAL BARRIERS

To begin with, there are physical barriers such as the skin. With the exception of a few openings the skin is a closed whole that protects everything inside it. The skin has little holes, or pores, but in principle things only come out of these: sweat and sebum. Sebum is fatty and protects the skin against water and most micro-organisms, at least if the skin has not been damaged. And that is the problem, because people are often unaware of small wounds. For us an opening of half a millimeter is hardly noticeable, but for bacteria—and certainly for viruses, which are much smaller still—0.5 millimeter is a sea of space.

Some "holes" in the skin are where they ought to be, such as the eyes, ears, nostrils, mouth, urethral opening, vagina, and anus, so that material can both enter and leave the body. Therefore the parts of the body that lie behind these openings are lined by a protective mucous membrane (mucosa). This mucous membrane protects us against dehydration, but it also has antibodies and can "imprison" disease-causing agents.

In the respiratory passages tiny hairs that sweep to and fro (cilia) work the agents of disease upwards, so that they can be coughed out. The mucosa of the stomach protects that organ against stomach acid, which is necessary for digestion but is also an obstacle for pathogens, just like the acidic environment of the vagina. And then there are also bacteria that are naturally present on the skin (or in the intestines) that keep disease-causing micro-organisms under control. These, too, are part of the defense.

WHITE BLOOD CELLS AND ANTIBODIES
When agents of disease penetrate farther into our bodies, via small wounds or in food, for example, they clash with the internal defense: the cellular and humoral (or specific) immunity. Cellular immunity comes from certain white blood cells: macrophages, granulocytes, and natural killer cells (a type of giant lymphocyte). These are cell-eaters (phagocytes) that clean up disease germs or afflicted body cells. They are not targeted at just one type of disease-causing agent. Therefore this is a form of non-specific defense. Yet there are cells that carry the antigens of a disease-causing agent on their own cell membranes, like warning lights. When these cells arrive in lymph nodes, the warning lights are shown to other white blood cells, and the specific defense can spring into action.

The specific defense (or humoral immunity) serves to fight well-defined disease-causing agents and consists of B- and T-cells (two other kinds of lymphocytes). T-helper cells activate B-cells and killer T-cells. T-cells fight viruses. They bind with an infected cell so as to kill it, ensuring that it cannot produce any more viruses. B-cells make antibodies that bind to the disease-causing agent. In this manner the disease-causing agent can no longer penetrate healthy cells (this is what happens to viruses), infected cells die directly, and they can easily be eaten up by phagocytes (this is what happens to bacteria). In the last case, so-called complementary factors also play an important role. The factors are proteins that, via a chain reaction, ensure microbes to which antibodies are attached are destroyed, among other means, through the production of trapping agents that attract the cell-eaters. You do not receive specific immunity at birth and must build it up during the course of your life.

DEMOCRATIC REPUBLIC OF THE CONGO (DRC)
Democratic Republic of the Congo (DRC) is the former Zaïre. Also called Congo-Kinshasa or just the Congo. Not to be confused with its much smaller neighbor Congo-Brazzaville, which is also called the Republic of the Congo. Generally, and also in this book, "the Congo" means the DRC. For the alternative country, the term Congo-Brazzaville is usually used. "Kongo" with a "K" refers to the River Kongo, which is the second longest river in Africa, after the Nile.

DEOXYRIBONUCLEIC ACID (DNA)

Also called deoxyribose nucleic acid. DNA is the most important carrier of hereditable information. A DNA molecule looks like a ladder that you have twisted to the left at the top and to the right at the bottom. The ladder's legs are chains of sugar-phosphate molecules, while the rungs of the ladder that bridge the gap between the legs are called base pairs. One base of each pair extends from a sugar-phosphate group to meet its opposite partner halfway along the rung. DNA is found in the chromosomes of a cell's nucleus, although there is also some in the mitochondria (the cell's power stations). It contains four different building blocks: the bases adenine, thymine, guanine, and cytosine, often shortened to A, T, G, and C. A forms a base pair with T, and G pairs with C. The order of the pairs in a string is called a sequence. There are very many possible sequences. In this way many different hereditable characteristics can be encoded. One or more DNA sequences together form a gene. In a chromosome there are tens to hundreds of genes.

Each cell has a copy of the whole organism's DNA. The DNA controls the formation of proteins that can perform all sorts of biological functions inside and outside the cell. During reproduction by two parent organisms, hereditable characteristics of both parents are transmitted to the next generation. In each of the offspring this happens in a different way, except for identical twins, triplets, etc. Thus brothers and sisters often resemble each other and their parents, but also differ from them.

DISTILLATION

Distillation is the technique whereby two or more fluids are separated by evaporation. The technique is based on the difference between the boiling points of the substances. If two liquids are mixed and you gradually heat the mixture, then the rates at which both substances evaporate increases. However, until the higher of the two boiling points is reached, the substance that has the lower boiling point will evaporate from the mixture more quickly. In the vapor that rises there is proportionately more of the substance with the lower boiling point than in the original mixture. By keeping the mixture at a temperature that produces a large difference between the rates of evaporation, and by maximizing subsequent capture of the vapor and letting it condense, you again obtain a liquid. It is possible to distill this liquid more times, until you have reached the desired purity. In this way you can, for example, obtain fresh water from salt water or distill alcohol from a mixture to make a stronger drink. At sea level the alcohol in drinks boils at 78°C. Or you can allow the steam from clean water to pass through herbs, whereby the water vapor carries away substances from the herbs such as organic oils. When you let the steam condense you obtain an herb distillate. African healers and herbal specialists sometimes use this technique to make all sorts of potions that have healing powers.

EBOLA MEDICINES

To be absolutely clear: at the moment no agents exist that have been shown to be safe and capable of curing Ebola. The standard treatment is still to let the patient rest, treat other problems (such as malaria), combat Ebola symptoms (with fever-reducing medications, clotting factors, etc.), prevent dehydration, and offer much TLC (tender loving care). The virus has been circulating for at least 39 years, but apart from the current monster epidemic, it remains a rare condition and moreover one that virtually exclusively strikes poor people in poor countries. That makes it too risky for the pharmaceutical sector to invest much money and time developing a therapeutic agent against Ebola. An advantage of the current epidemic is that there are now test subjects galore upon which possible experimental agents can be tested, by way of compassionate use. And we do have experimental agents, because fortunately during past years a few colleagues and businesses still invested in research on Ebola and Marburg medicines.

During the first Ebola outbreak in Zaïre in 1976, we ourselves had experimented by administering serum from convalescent EBOV patients to acutely sick EBOV patients. During the epidemic in Kikwit in the Congo nineteen years later, my ITM colleague Bob Colebunders did the same. However, there is not yet any convincing evidence that the administration of sera with EBOV antibodies can really cure people. In 2015 a project started in Guinea under the leadership of Dr. Johan van Griensven of the ITM, to evaluate the effect of immune serum on patients with Ebola. In 2016 the study showed that the administration of serum containing EBOV antibodies did not cure such patients.

Another experimental medicine, TKM-Ebola, goes to the heart of the matter. Marburg and Ebola are filoviruses that are made from an RNA string in a protein capsule. Thereby they can invade cells and hold the protein factories of the cell hostage in order to make copies of themselves. About ten years ago filovirologist Thomas Geisbert of the University of Texas found a way to stop the multiplication of filoviruses by administering a small interfering RNA molecule (siRNA) to infected test animals. The molecule does indeed interfere, because it binds itself to the genes of the virus and so makes them harmless. Normally the body breaks down such molecules, but the Canadian company Tekmira thought of a way to pack them in microscopically small capsules, so-called fat nanoparticles, whereby they still enter the infected cells. Thus researchers from Texas and Tekmira could work together to develop siRNA medicines against different strains of Marburg and EBOV (Zaïre Ebola virus, the virus that caused the 2013–2015 epidemic in West Africa). From experiments with guinea pigs and monkeys it appears that treated animals can survive doses of the virus that are usually fatal, on condition that they receive the medicine within one or two days after they have been infected. The problem is that most infected people have no symptoms that early, have not yet sought medical aid, and the virus itself cannot yet be detected in the blood. But in August 2014 a new study of monkey experiments appeared in the journal *Science Translational Medicine*. From this work it appeared that the medicine works even if it is only administered three days after the animals are infected. The extra day makes a world of difference, because from then on the experimental animals do have symptoms. The important news was that a limited clinical study has already occurred with a promising result, although it has yet to be seen whether it is also valid for people with Ebola, without too many side effects, and indeed if the method is practical in an African outbreak. The monkeys received the medicine each day for a week, by infusion.

The administration of monoclonal antibodies to rhesus monkeys in the form of the experimental ZMapp also led to the complete recovery of all experimental animals. During the current outbreak ZMapp was also tested on people, with mixed results, but the number of infected patients was too small to be able to say anything about the efficacy of the medicine. The supply of ZMapp was too small for it to be tested on a large scale. It is now being produced in much larger quantities, in collaboration with businesses that manufacture ZMapp in their own plants.

TKM-Ebola and ZMapp are only examples. To date, the abilities of some fifteen agents to inhibit the growth of Ebola and Marburg in cell cultures, and sometimes also in infected mice, have been tested with success. In this regard testing is difficult because you must work with the live Ebola virus, which can only be done in a BSL-4 lab. A quicker alternative would be to test the antiviral efficacy of all such agents on the Vesicular Stomatitis Virus (VSV). Among other species this virus strikes horses and cattle, but has no serious consequences for people. Thus you do not need a maximum security lab to work with it, while the way that VSV multiplies inside cells strongly resembles the method that is used by Ebola. This line of attack is now being followed.

EBOLA TESTS

An outbreak and certainly an epidemic can only be stopped if enough simple and reliable laboratory tests are available. Only then can you confirm or rule out suspected cases sufficiently quickly. This is necessary because you can't quarantine everyone and keep them there. Without such a test it is hard to know for certain if someone is infected, especially if the infection is only a few days old. The first symptoms of Ebola are always atypical, and could just as well indicate malaria or typhus, for example. Other diseases with which Ebola can be confused at the beginning are dengue, cholera, yellow fever, measles, hepatitis, sleeping sickness, and a host of other tropical afflictions. But non-infectious problems such as snakebites and clotting diseases can at first look like Ebola. Moreover, patients in third world countries with bad health care often suffer from multiple ailments simultaneously. Someone can have both Ebola and malaria. To confirm malaria is thus no way to rule out Ebola.

In step with the progression of the disease, the symptoms increasingly begin to indicate hemorrhagic fever, but different forms of symptoms exist that strongly resemble each other but have different causes and demand a different attack. Thus the serum from a cured Marburg patient does not help to fight Ebola, notwithstanding the fact that the viruses are closely related. This we ourselves noted in Kinshasa in 1976.

Fever on its own shouldn't ring any alarm bells, but if someone has also had contact with the blood or other body fluids of a (possible) Ebola patient (human or animal), or with objects that could be contaminated with such a fluid, then the person in question must be isolated and the health services informed. Thus samples can be taken from the patient for testing in the lab to establish if it is indeed an Ebola infection.

To begin with, there are NON-SPECIFIC METHODS to obtain information about someone's Ebola status by laboratory investigation. Warning signs are too few blood platelets (thrombocytopenia), the number of white blood cells in the blood is at first low and thereafter high, raised liver enzymes, and abnormal blood clotting parameters. THE VIRUS ITSELF CAN ONLY BE DETECTED WHEN THERE ARE SYMPTOMS. This is one of the problems in outbreaks: you must keep people who are possibly infected close at hand, under safe conditions. And even after the first symptoms begin it can be days before there is enough virus in the blood for it to be detectable.

That detection can be done with specific laboratory techniques. PCR is probably the most precise technique, as well as the earliest detector, although the test can be negative in an infected person during the first three days after the first symptoms appear.

In an outbreak, VIRUS ISOLATION is often unsuitable as a detection method. In virus isolation you culture the virus and thereafter search for it with the electron microscope, as we did in 1976. That takes a long time and is dangerous, so that this method can only be applied in a maximum security lab, and there are only a few tens of these labs in the whole world.

In field hospitals and mobile laboratories, the most used and sensitive tests are RT-PCR tests. But PCR is finicky; it requires well-trained staff and a full vial of blood to do it, and just one test costs 70 euros. These are not simple matters in West Africa. If, on top of this, you must transport blood samples to a limited number of laboratories over bad roads, much time is lost. As a precautionary measure people suspected of having the disease must remain in transit centers during this time, which they might enter with just malaria, but while expecting their test result they leave with Ebola, and leave between six planks if it takes too long.

During the West African outbreak of 2013–2015 the lack of fast and simple tests was an enormous problem, at least in the first twelve months. In Liberia new mobile test equipment could deliver reliable test results in three to five hours. But even that is too long.

The ideal test, so the WHO has specified, must be capable of being performed in remote and temporary health centers without laboratory infrastructure, must comprise only three steps and must deliver a reliable result within half an hour — and all that must be possible without any other safety measures besides personal protection. Moreover the substances that you need for such an ideal test must be easily stored and mixed, and it must be possible to train the lab technicians to do the test in less than half a day. The apparatus must be able to work without an external energy supply and require virtually no maintenance.

It seems too good to be true, but since December 2014 different systems have been tried in West Africa that approach this ideal, such as a mobile "coffer" laboratory that is powered by a solar panel and uses compressed reagents in a pellet form that requires no cooling. It delivers a result within quarter of an hour, on the basis of genetic material in the virus—thus it is specific and sensitive.

In Belgium Ebola tests are undertaken at the Institute for Tropical Medicine in Antwerp. Until the end of October 2014, suspect samples had to be sent abroad, which extended the waiting time by 12 to 24 hours. But now the tests can be done in the ITM's own laboratories. The ITM already had the technical capacity, but to test for Ebola itself the institute required the permission of the government, and that fiat has now been arranged. Under the leadership of my successor Kevin Ariën, a blood sample is first neutralized in a high security BSL-3+ laboratory (the highest security level available in Belgium), and thereafter Ebola RNA may be isolated and investigated with advanced molecular diagnostics. Ebola cannot be cultured and studied as we did in 1976, because a maximum security BSL-4 lab is required for this.

See: maximum security lab, vero cells, virus test

EBOLA VACCINES

At the time of completing this book, I know of eleven vaccines that are under development. Most are therapeutic vaccines that are thus intended for use in people who are already infected with Ebola. Such a vaccine will prevent death and accelerate healing. In the best case it will lead to the virus being driven out of the body and the vaccinated person being better protected against new EBOV infections. The advantage of a therapeutic vaccine is that it only needs to be administered once to offer protection for a specified period (weeks, months, years). By contrast, antiviral treatment agents must be swallowed or injected daily.

Work is also being done to develop a preventive vaccine that would stop the virus invading the body and—if it should gain entry—prevent it multiplying. The pharmaceutical company GlaxoSmithKline is working together with the American National Institute of Allergy and Infectious Diseases (NIAID) on such a preventive vaccine. In a preparative study it was shown that all sixteen monkeys to which the vaccine was administered were protected against EBOV. Meanwhile clinical studies on people have begun in the United Kingdom and the United States and will be continued in Mali and Switzerland. The goal is to vaccinate 260 participants. In this phase of the clinical studies, it is being investigated whether the vaccine causes side effects in healthy volunteers and whether it stimulates an immunological response. The first results in the US give hope: the GSK-NIAID vaccine is well-tolerated and delivered an immunological response in each of the twenty volunteers who received injections.

You may well ask yourself why the GSK-NIAID vaccine is suddenly the subject of clinical evaluations. NIAID had already started to develop an Ebola vaccine in 1990 so that the United States would be able to defend itself in the event of a bioterror attack.

A second vaccine is being developed by the Public Health Agency of Canada together with the American firm NewLink Pharmaceuticals. It appeared to protect twenty monkeys infected with the Ebola virus. The license for this vaccine has now been transferred to the pharmaceutical company Merck. The first phase of the clinical studies began in October 2014 in the US. A month later volunteers were also vaccinated in Gabon, Germany, Kenya, and Switzerland.

More than 200,000 units of a third vaccine developed by Johnson & Johnson have already been produced, with an eye toward large-scale clinical studies. In the course of 2015, production of this vaccine will be raised to two million units. The vaccine was developed by Johnson & Johnson in close collaboration with its subsidiary Crucell Holland BV and Bavarian Nordic, a Danish biotech business. The vaccine is constructed from an adenovirus vector into which a piece of genetic information has been inserted, this piece being derived from the Ebola virus that is currently circulating in West Africa. After a first injection (into the muscles) a second follows a little later. The development of this vaccine is part of a larger study, in which a preventive supervaccine is being sought that will offer protection against all species of Ebola and Marburg viruses. In animals the results of this work are hopeful.

In the US the firms Profectus Biosciences, Protein Sciences, Novavax, Vaxart, and Inovio have begun to develop their own vaccine. Work is also being done on Ebola vaccines in Russia. If the number of people newly infected with Ebola continues to decrease, it will probably become impossible to assess the efficacy of the different vaccines. Hopefully the pharmaceutical industry will still carry out phase 1 and phase 2 of development (the clinical trials), so that the side effects and safety of the vaccines can be precisely mapped, and so that the optimal quantity of vaccine required to stimulate an immune response that is as strong as possible becomes known. In this way we will be able to start vaccinating much sooner during a new large-scale epidemic and will be able to measure the efficacy of the vaccine. People active in the health sector will be first to receive vaccines. That will allow them to care for Ebola patients much more efficiently without being hindered by uncomfortable protective clothing in which they rapidly become overheated. In this manner vaccinated aid workers will also scare patients less, communicate more easily, and generally build greater trust.

EBOLA VIRUS

The Ebola virus is a genus of RNA viruses within the family of filoviruses (Filoviridae), and is related to the Marburg virus. In monkeys, apes, and people Ebola often—but not always—causes serious internal bleeding. The incubation period lasts from two to twenty-one days. The first symptoms are usually vague: fever, severe headache, an aching throat, and a general feeling of being unwell. Very quickly it becomes worse and muscle pain, loss of appetite, and intestinal complaints are added. Sometimes a rash appears after roughly a week, usually on the buttocks. In about 50 percent of patients the mucosa begin to bleed, which is visible in their vomit and feces. In the second week people die, or begin to recover, usually slowly.

The Ebola virus is spread by body fluids and often kills entire villages. Thereafter it once again disappears to slumber somewhere in the natural world—possibly in fruit-eating bats.

There are five species in the genus Ebola. The deadliest species is the virus that circulated in 1976 in Yambuku. First is the Zaïre Ebola virus, or more simply the Ebola virus (presently abbreviated as EBOV). This is also the virus that caused the recent outbreak in West Africa, the greatest in history. A few months before Yambuku 1976, another outbreak of a hemorrhagic fever virus occurred in the southern Sudan. From this outbreak our British colleagues Bowen, Lloyd, and Simpson isolated the second species, the Sudan Ebola virus (now abbreviated to SUDV), in the maximum security laboratory at Porton Down in Great Britain. At the CDC in Atlanta, Patricia Webb allowed serum with antibodies against EBOV to react with SUDV and only observed a very weak reaction. Thus EBOV differed from SUDV. Many years later this was substantiated by genetic research. The three other Ebola species are the Bundibugyo Ebola virus (BDBV), the Reston Ebola virus (RESTV), and the Taï Forest Ebola virus (TAFV).

EBOV

Zaïre Ebola virus, the virus of Yambuku 1976 and West Africa 2013–2015.
See: Ebola virus

ELISA

See: virus test

EPIDEMIC

Derived from the Greek word epidémios, which means "over the entire population." Usually the concept of an epidemic refers to an infectious disease that is present to a conspicuously greater degree than normal. The opposite of an epidemic is an endemic or endemic disease, which is continuously present but in fewer people. A pandemic is an epidemic in many countries or world regions. The Spanish Flu of 1918, which killed about twenty million people worldwide, was a pandemic. There have been countless epidemics in history. Often they were closely tied to famine, not to mention war, because malnutrition greatly weakens the body's defenses. Some examples:

- In the second century millions of people died in the Roman Empire, probably from smallpox.
- In the sixth century the Byzantine Empire was struck by the plague, which also gripped other parts of Europe and possibly cost the lives of a quarter of all Europeans.
- In an even greater plague epidemic in the middle of the fourteenth century, the Black Death claimed the lives of about 25 million Europeans from a total population of 85 million.
- After Columbus discovered America in 1492, a large part of the population of the New World—in some areas up to 90 percent—was killed by imported diseases to which the Native Americans were not immune. The greatest devastation was inflicted by smallpox.
- In the seventeenth century the plague struck Western Europe once again, killing 60,000 in London and 34,000 in Amsterdam.
- Straight after the first World War there was the Spanish Flu, which claimed twice as many victims as the war itself.
- A recent example is AIDS, a pandemic that is especially at home in different African countries, with almost 80 million infections worldwide and nearly 40 million deaths during the past 35 years.

Many countries have a policy to prevent and combat epidemics. In this regard enlightenment through education and the media is of great importance. This can be about vaccination and other precautionary measures such as washing hands and safe sex. If an epidemic nevertheless breaks out or the threat is significant, the government can then limit people's freedom of movement by closing airports and borders, placing suspected cases in quarantine (isolation, if necessary by force), and declaring a state of emergency. Epidemics do not only claim victims directly, but can also have extremely serious economic consequences because people are no longer working: they are ill, may no longer leave a specified area, or are scared to go to work. They also spend far less, and if continued, it can bring transport and the supply of goods to a standstill. If this happens to an economy already severely hampered by corruption, civil war, or famine, a country can enter a vicious circle in which even elementary health care is absent and new epidemics regularly break out.

During epidemics that spread rapidly through densely populated areas, people try to flee. The chaos that can arise from this sometimes descends into violence. This helps the disease to spread farther and makes it difficult to detect sick and (possibly) infected people. Moreover, refugees often end up in even worse circumstances, in which new epidemics can break out.

Epidemiology is the science of studying the spread of diseases and the prevention thereof. Thus an epidemiologist tries to find who has been struck by the disease and the contributing factors: age, gender, places visited, diet, etc. They attempt to map the epidemic, to find the source of infection and so ring-fence its spread and stop it. A famous example from history is a British doctor who fought a cholera epidemic in the London district of Soho in 1854 by questioning patients. That led him to a local water pump. He removed its handle and the epidemic died out. The cholera bacterium was itself only discovered later, but the doctor had indeed found the source of infection and, in doing so, laid the foundation of modern epidemiology.

See: index case, outbreak

FILOVIRUSES

Filoviridae, a family of threadlike viruses. Since the first documented outbreak of the Marburg virus (MARV)—a description of this is provided in Chapter 2 under "August– September 1967"—only two other genera of filoviruses have been discovered. The most recent representative is the Lloviuvirus (LLOV), which was first mentioned in 2010. The Lloviuvirus belongs to the genus Cuevavirus, after the Spanish cave Cueva del Lloviu where it was discovered.

And then, of course, there is that other filovirus genus: Ebola. Outwardly Ebola strongly resembles Marburg, gives rise to similar symptoms, and is approximately equally deadly. It is highly probable that it shelters in specific bat colonies, just like Marburg. No wonder that in 1976, when we discovered Ebola, we at first thought it was Marburg virus. An important difference is that Ebola appears more often, and from December 2013 would even lead to an epidemic with about ten thousand fatalities. After 1967 there have been seven outbreaks of Marburg, but even the deadliest of these—that at the end of 2004 in Angola—cost "only" 356 human lives. The only Marburg infection to reach the major media of the Netherlands was that of the 41-year-old Astrid Joosten. In the summer of 2008 she was on holiday in Uganda, and inside a cave she made contact with a bat or bat droppings, became seriously ill after returning home to the Netherlands, and shortly after died in Leiden.

FOMÉTRO (FONDS MÉDICAL TROPICAL)

Medical Tropical Foundation, established by medical faculty members of Belgian universities

HAART

Highly Active Anti-Retroviral Therapy

See: AIDS

HANTAVIRUS

Hantaviruses belong to the family of Bunyaviruses (*Bunyaviridae*) and are transmitted by diverse species of rodents. About twenty species of hantaviruses exist, of which roughly half can make people ill. The symptoms are hemorrhagic fever, low blood pressure, and kidney problems. The course of the disease in a patient can be mild, but in some cases they die quickly. Another dangerous form causes heart problems and hinders breathing.

The hantavirus was first spoken of during the Korean War of 1951 to 1953. Then, three thousand UN soldiers fell ill from the disease, of whom 10 percent did not survive.

Just like Ebola and Marburg, hantaviruses are single-string RNA viruses. You become infected by breathing in air that contains virus particles that have been shed by rodents. The incubation period is one to six weeks. Not everyone who is infected falls ill.

The hantavirus known as Puumala is present in the Low Countries. Its symptoms are fever, cholic, vomiting, bleeding of the eye's membranes, headache, and a slow heartbeat. The Puumala virus is transmitted by the bank vole (*Clethrionomys glareolus*). In Flanders this species is present in great numbers in beech forests with undergrowth, poplar forests, pine forests, forestry plantations, reed beds, parks, and gardens. The bank vole likes moist soil and a thick layer of humus. They can climb well and are encountered meters from the ground in trees.

HEMORRHAGIC FEVER

"Hemorrhagic" comes from the Greek word *aimorragia*, which refers to prolific bleeding. Thus, in brief, hemorrhagic fever is a sickness that causes both fever and bleeding. Fever is always present, but not always bleeding. Most forms of hemorrhagic fever are caused by a virus. The best known are Ebola fever, Marburg fever, Lassa fever, and yellow fever. They are all relatively rare and foreign to Europe, but are serious diseases. From 10 to 80 percent of patients die.

The disease breaks out three to twenty-one days after infection with flu-like symptoms such as fever, muscle ache, abdominal pain, joint pain, and vomiting. In this phase the disease is hard to recognize. Following this the patient either recovers spontaneously or the situation rapidly worsens and hemorrhage begins: under the skin and in organs, and/or from the nose, mouth, eyes, ears, and anus. Sometimes there is multiple organ failure (MOF) in which several organs fail at the same time.

Belgian doctors who come into contact with a suspected patient must report this to the government on a special form. Because hemorrhagic fever viruses are spread by animals, the disease begins in places where such contact occurs. This happens during the butchering and eating of animals, through bites, or contact with droppings (often when feeding livestock). Transmission from person to person happens through the contact of damaged skin or mucosal tissues with blood and other body fluids. Airborne infection has never been shown to occur in humans.

HIV

Human immunodeficiency virus
   See: AIDS

HUMAN GENOME PROJECT

The human genome project is funded by the American government with the goal of identifying and locating all human genes. Other countries are also taking part. When the project was launched in 1988 the target seemed unattainable, but science subsequently evolved so rapidly that already by 2000—five years earlier than planned—a map of the human genome existed. Three years later 99 percent of the human genome was known with 99.99 percent accuracy. Alongside this scientific development there was also a concurrent project by Celera Genomics, which sped up the research. Since then there have been all kinds of similar initiatives to unravel the DNA of specific organisms. Thanks to the human genome project, we know the three billion base pairs of the human genome. Many genes are also known, although that part of the research is ongoing. We also know now that within human DNA there are "only" 20,000 to 25,000 genes, about half the number originally estimated. The knowledge delivered by the

HGP is freely accessible on the Internet, although you need bespoke software to work with the information.

IMMUNE SYSTEM
See: defense

IMMUNITY
See: defense

IMMUNIZE/IMMUNIZATION
To make immune is to make a person invulnerable to a disease-causing agent. There is active and passive immunization. In passive immunization you give people antibodies from other people or animals. That has a quick but short-lasting effect. In active immunization immunity is generated by exposing a person's immune system to an antigen or antigens, such that the person produces their own immunity. That offers long lasting, possibly even permanent, protection.

Active immunization can occur naturally when a microbe or antigen invades the body. The immune system then makes antibodies against the micro-organism, but that goes slowly and is not without risk. In fact the micro-organism can make the patient ill or kill them. Therefore artificial active immunization also exists. In this case a weakened or killed micro-organism, or a piece thereof, is injected into the patient. Most vaccinations that children receive are a form of artificial active immunization.

INCUBATION PERIOD
Or incubation time. The time that passes between the infection and the first symptoms. For many infectious diseases this is one to three weeks, but an incubation period can also be significantly shorter. For example, in cholera it can sometimes amount to just six hours, and you can already suffer from the flu within 24 hours of infection. Or equally, the incubation period can sometimes be much longer: if you contract the Epstein-Barr virus (EBV) it can take three months before you develop glandular fever. HIV is absolutely extreme; on average it is only nine to ten years after infection that it causes the first symptoms of AIDS if the virus is not inhibited.

INDEX CASE
The index case is the starting patient in the population (the group of people who are being investigated). Occasionally "index patient" is intended to mean the first case of a disease that has been described in the medical literature, regardless of whether the patient was the first to have the illness. The use of the term "index case" can indicate that an outbreak is being discussed, the source of the disease, the possible means by which the disease spread, and the nature of the reservoir in which the disease shelters between outbreaks. The index case is sometimes also called "patient zero."
See: epidemic, outbreak

INFECTION
An infection happens when a micro-organism, a virus, a prion (a disease-causing abnormal protein), or a parasite enters a living being and also multiplies within it. The

latter is important, because without multiplication what you have is a CONTAMINA-TION, not an infection. True infections are often damaging, but need not be so. Beneficial intestinal bacteria also multiply and are not damaging—quite the opposite. You even need this sort of infection.

Only if an infection inflicts so much damage that you can no longer function normally and/or painlessly do you have an INFECTIOUS DISEASE. An infection often causes INFLAMMATION. Therefore the concepts of an "infection" and an "inflammation" are often used for each other. But an inflammation is something else. An inflammation is a reaction of your body against tissue damage or external provocation. Inflammation is useful, because through it damaging substances are removed or neutralized and possible damage is repaired. An infection often causes inflammation, and vice versa an inflammation can give rise to an infection, but the two can also appear independently. You can get an inflammation from a micro-organism or another disease-causing agent that has entered your body and multiplied, but also from sitting in the sun too long, from a bruising, from exposure to a poison, pollen, gluten, or another allergen that you cannot tolerate. Auto-immune diseases such as rheumatism and Sjögren's syndrome also bring about inflammatory reactions.

INFECTIOUS DISEASE
See: infection

INFLAMMATION
See: infection

INSTITUT PASTEUR
The Institut Pasteur is a French private, non-profit organization for the study of micro-organisms, diseases, and vaccines. The institute was opened in 1887 and named after its founder, the French scientist Louis Pasteur, who in 1885 developed the first vaccine against rabies. Pasteur had earlier undertaken research for breweries and conceived a preservation technique that would be known as pasteurization. His establishment grew into a leading international research institution with its headquarters in Paris. The Institut Pasteur has contributed to many of the discoveries that have been made in the fight against infectious diseases such as tetanus, tuberculosis, polio, influenza, yellow fever, the plague, and AIDS. Eight scientists of the institute have received a Nobel Prize for medicine or physiology. The Institut Pasteur is among the best research institutions in the world. It has a hundred research units, with more than one thousand scientists from seventy countries. The organization has an international network of twenty-four Pasteur institutes outside France, many in developing countries.

INSTITUTE FOR TROPICAL MEDICINE (ITM)
The Prince Leopold Institute for Tropical Medicine (in Flemish: Prins Leopold Instituut voor Tropische Geneeskunde)—known locally as the "tropical institute"—is a research establishment in Antwerp for tropical medicine and health organization in developing countries. The ITM was founded in Brussels in 1906 as a school for tropical medicine, above all else in response to the needs of the Congo Free State, the personal colony of Leopold II. In 1933 the institute moved to the port city of Antwerp. During the course of

the twentieth century the ITM became famous for its research on HIV, malaria, tuberculosis, and "neglected tropical diseases." The ITM also gives health care advice and vaccinations to travelers, treats people with a tropical disease in Belgium, and annually trains about five hundred doctors, veterinarians, biomedical staff, and nurses from all over the world.

Among others, in the 1980s Dr. Peter Piot of the ITM showed that AIDS did not originate among homosexuals on the American West Coast, but that it is a tropical disease that chiefly strikes African heterosexuals. AIDS and HIV remain priorities for the ITM.

THE ITM AND EBOLA

Since the ITM is the Belgian reference center for tropical and infectious diseases, it is also the main reference center for Ebola. Before the Ebola epidemic of 2013–2015, the ITM infectious diseases expert Dr. Erika Vlieghe, in her capacity as the national Ebola coordinator, advised the federal Minister of Public Health, Maggie De Block. Erika Vlieghe manages the Ebola coordination team, coordinates Belgium's contribution to the global attack on Ebola, and coordinates the flow of information to health professionals and the public. The ITM also develops the procedures, training, and infrastructure needed to receive Ebola cases at the ITM and the University Hospital of Antwerp. The institute collaborates in formulating the guidance offered by the High Council for Health (Hoge Raad voor Gezondheid) and the World Health Organization. Furthermore, the ITM is ready to diagnose Ebola in one of the high security BSL-3 labs at the institute. Lastly, the ITM answers countless questions from colleagues, organizations, and the media.

The ITM is also internationally active in the fight against Ebola. Thus the institute leads an international consortium that is investigating whether infected patients in West Africa can be treated with antibodies from the blood of people who have survived Ebola. Moreover the ITM assists the initiatives of Doctors Without Borders, the European Mobile Laboratory Project, and the Global Outbreak Alert and Response Network of the World Health Organization. The ITM also advises the federal and Flemish governments with regard to the epidemic in West Africa.

LASSA VIRUS

Each year the Lassa virus contaminates a few hundred thousand people in West and Central Africa. They encounter it through contact with the droppings or urine of contaminated animals, for example in places where grain is stored. The host of the virus is the multimammate mouse, a type of rodent that is widely present south of the Sahara and that can transmit other diseases, including the plague. Although an infection with Lassa usually passes without problems, about five thousand infected people die each year. Many contaminated people have antibodies and therefore do not know that they have ever had the virus. The antiviral drug ribavirine can save lives, if it is administered early by infusion.

See: hemorrhagic fever

LEUCOCYTE

White blood cell.
See: blood

LLOVIU VIRUS
See: filoviruses

MALARIA
Malaria is derived from the Latin "mala aria," which means "bad air." This refers to the stench in the vicinity of swamps. Malaria is also sometimes called swamp fever. However, the disease has nothing to do with bad air but rather Plasmodium falciparum, a parasite that is transmitted to people by mosquitoes. This primarily occurs near still-standing water and swamps, hence the confusion.

ROUGHLY SIX HUNDRED THOUSAND PEOPLE DIE OF MALARIA EACH YEAR, ESPECIALLY YOUNG CHILDREN IN AFRICA. Until 1938 the disease also occurred in Belgium. Through land drainage, the use of artificial fertilizer and pesticides, and the erection of better toilets, the disease was as good as eradicated. Extremely rarely, malaria still crops up in Belgium as the result of a local cause, but most domestic cases arise in returning tourists or immigrants from the tropics. In any case, the fact that so many malaria deaths occur in Africa has more to do with poverty than climate. There are different species of plasmodium which can give people malaria, but Plasmodium falciparum causes by far the most cases of this disease, and also the most deadly.

SYMPTOMS
The symptoms of malaria are bouts of fever with cold shivers and sometimes vomiting. In accord with the species of plasmodium, the bouts occur every 48 to 72 hours, but can also be more frequent and irregular, as in Plasmodium falciparum. In this most dangerous form blood vessels often become blocked. If that happens in the brain, the patient has cerebral malaria. Without correct treatment the risk of death then rises to almost 100 percent. Even with good treatment 15 to 20% of patients with cerebral malaria die. I have myself contracted malaria and carry the plasmodium parasite in my blood, but it hasn't played up for many years.

IMMUNITY
In many African countries, much of the population is contaminated from a young age. With time, children who survive the initial infection build some defense against the disease. But because people are also traveling more, malaria variants are also spreading more quickly and this tends to diminish the immunity of local populations.

ATTACK AND TREATMENT
Malaria can be tackled if the sickness is discovered and treated early enough. Diagnosis requires a drop of the patient's blood. The technique is simple, because you only need dyes and a microscope.

Different drugs are available and new ones are regularly added. This is essential, because the parasites easily develop resistance. People who do not live in the tropics but are temporary visitors there can take preventive medicines. A vaccine still doesn't exist, but work is underway.

Malaria cannot be transmitted from person to person. An intermediate step is required: the anopheles mosquito. In the Low Countries these mosquitoes are barely present and so the chance that you will contract malaria or transmit it through a mosquito is extremely small. People who have recovered from malaria can have normal contact with others and go to work or school.

In large parts of the tropics malaria-carrying mosquitoes are present in overwhelming numbers, and the risk that you will be contaminated is very real. Prevention remains best, and thus, above all else, the malaria-carrying mosquitoes must be fought. That can be done with mosquito nets that have been dipped in an insecticide or a repellant such as DEET and by draining away still-standing water. Research is also being done on genetically modified and sterile mosquitoes.

MARBURG (OR MARV)
Marburgvirus, a genus of filoviruses.
See: filoviruses

MATABICHE
A fee, usually intended as a bribe. Comparable with baksheesh, a concept that is encountered more in the Middle East and Asia. Matabiche is a word from colonial times, heavily used in the Belgian Congo and French Equatorial Africa. It comes from the Portuguese "matabicho," which in turn is a contraction of "matar o bicho," or "to kill the worm." This you do by drinking a glass of wine or another alcoholic drink on an empty stomach. Matabiche evolved from this to mean something that you take or give to solve a little problem.

MAXIMUM SECURITY LABORATORY (MSL)
A maximum security laboratory is a lab of the type BSL-4. BSL stands for "biosafety level," a term that indicates the conditions that must be fulfilled in a room in which work is being done with biological materials such as micro-organisms (viruses, molds, bacteria, parasites, protozoa) or biological poisons (toxins). The conditions serve not only to protect personnel but also the world outside the lab against accidents, infection, and spreading of the biological material.

- BSL-1: Here work may only be done with biological materials that normally cannot make any healthy people sick and that present little risk to laboratory personnel. Normal laboratory hygiene together with the use of hand gloves and a form of facial protection suffices. Waste is disinfected. The lab does not need to be separated from the general circulation of people in the building. Contaminated materials normally go into normal but well-marked containers.

- BSL-2: Here work may also be done with biological materials that can cause sickness in people but cannot spread easily through the population (e.g. cannot be transmitted through the air), and for which effective treatment agents or vaccines exist: whooping cough, diphtheria, measles, meningococcus, etc. The only people who may enter the lab are the people who work there and know the procedures. The doors must always be closed and the windows cannot open. The room can be thoroughly cleaned and disinfected. The personnel have received specific training in the handling of the disease-causing agents and are supervised by scientists. Access to the lab is limited when work is being done there, strong precautionary measures are enforced with regard to contaminated sharp objects, and work activities during which contamination could occur, via the air or spatter, take place in a closed container, such as a biological safety cabinet.

- BSL-3: Here work may also be done with biological materials that can cause very serious and/or deadly sickness in people and can spread, but for which effective treatment agents or vaccines are available: AIDS, polio, tuberculosis, plague, SARS, typhus, yellow fever, rabies, etc. No contaminated air can escape from a BSL-3 lab because the air pressure inside it is always lower than that outside and because the air removed by the air conditioning system is filtered. You can only enter by means of an airlock and in a special overcoat that closes at the back. Work is done in a biological safety cabinet or a glovebox. Materials taken out of the lab must be double-packed and decontaminated.

- BSL-4: Here work may also be done with organisms that are contagious and deadly, and for which no effective treatment agents or vaccines exist, such as the Ebola, Marburg, and Lassa viruses. (Smallpox is an exception—vaccines do exist, but use of a BSL-4 lab is still required.) This lab is a hermetically sealed space within seam-welded stainless steel plates. Work occurs in isolation boxes and the personnel wear pressurized suits that have their own independent air supply in which no air from outside can enter. There are different showers, a vacuum chamber, an ultraviolet chamber, electronically secured airlocks, and other precautionary measures to neutralize all possible traces of dangerous materials. All air and water that enters or leaves a BSL-4 lab is disinfected. Access to the lab is strictly controlled by the person in charge. The lab is located in its own separate building or within a fully closed section of a larger building.

MICRO-ORGANISM
An organism that is so small that you cannot see it with the naked eye. These are bacteria, protozoa (such as amoebas and spores, including the cause of malaria—plasmodium), single-cell algae, and molds (including yeasts). Viruses and prions are not living, are not organisms, and hence are not micro-organisms.

MONOCLONAL ANTIBODY
A class of antibodies that is manufactured in the laboratory and with which specific diagnostic tests can be made, among other things.

WHAT ARE ANTIBODIES AND ANTIGENS?
Antibodies are proteins that are produced by a specific type of white blood cell of our defense system (B-cells) as a reaction to other proteins that are alien to the body. We call the latter proteins "antigens." It is by their antigens that viruses and bacteria are recognized in the body of a patient and combatted. The defense reaction occurs through an-

tibodies that attach or bind themselves to the undesirable antigens. That doesn't take place all over an antigen, but only at specific parts of it. We call these little parts "epitopes."

POLYCLONAL IN NATURE
In nature antibodies are polyclonal. That means that they come from multiple ("poly") clones of B-cells. Due to this they secure themselves to multiple epitopes. That is favorable for defense. But at some time, researchers thought of extracting antibodies from the blood of people or animals that have survived a disease. They wished to use the antibodies for immunization and thereby prevent others falling ill, or to give the defense of people who were already sick a little push in the the right direction. Moreover, from the antibodies of a contaminated or sick person you can tell what disease-causing agent is responsible.

MONOCLONAL FROM THE LAB
If you inject ordinary polyclonal antibodies into someone to immunize or treat them, you run the risk that there will be dangerous side effects. And polyclonal antibodies are equally poor for the diagnosis of diseases. The problem is that they are not always sufficiently specific. You will indeed know that a certain class of pathogen is involved, but you will not always know precisely which type.

In 1975 scientists devised a technique to make monoclonal antibodies (or mAbs) in a lab. MAbs have the advantage that they only bind with a single epitope of a specific antigen. That makes them highly suitable for the diagnosis and combatting of diseases. To make a mAb laboratory mice are injected with antigens that have the epitope in which you are interested. Thereafter the spleen of such a mouse is removed and the B-cells are extracted from it. The cells are mixed with myelomas, a type of cancer cell that multiplies continuously. If a B-cell fuses with such a cancer cell, we obtain a new cell type that is called a hybridoma. These hybridomas are immortal and continually produce the same antibody.

THE IMPORTANCE OF THIS
As soon as you have monoclonal antibodies against a specific epitope, you can also use them to detect the presence of a substance with that epitope inside an organism. Therefore monoclonal antibodies are of great importance in the development of tests to find a specific pathogen.

MSL
See: maximum security laboratory

MYCOBACTERIA
See: tuberculosis

NCD
Non-communicable disease
   See: non-transmissible disease

NIAID

National Institute of Allergy and Infectious Diseases. An American research institution in Bethesda, Maryland. NIAID is one of the 27 establishments and centers that together form the National Institutes of Health (NIH), an agency of the American Public Health Department. NIAID's mission is to undertake fundamental and applied research on infectious diseases, diseases of the immune system, and allergies, in order to understand, treat, and prevent them.

See: Ebola vaccines

NON-TRANSMISSIBLE DISEASE (NTD)

A disease that is not contagious and cannot be transmitted from one person to another. By this we not only mean chronic, slowly progressing diseases but also diseases that can in some cases kill rapidly. The duration of the disease plays no role, but rather the fact that they are not caused by an infection. Examples of such diseases are forms of auto-immune diseases, heart diseases, strokes, cancer, asthma, diabetes, chronic kidney disease, chronic lung disease (COPD), osteoporosis, Alzheimer's, cataracts, obesity, and many other conditions. According to the World Health Organization non-transmissible diseases are responsible for 60 percent of all deaths, and are thus the most important causes of death in the world.

NUTRITIONAL SUBSTRATE

A nutritional substrate or growth medium is a solid or fluid substance in which micro-organisms, plants, or animals can be cultivated. For the investigation of micro-organisms such as molds and bacteria, some nutritional medium is poured into a petri dish (a small, flat, round dish of glass or plastic). The medium consists of nutrients that have been dissolved in agar, a gelatinous substance extracted from dried seaweed. Much used nutritional mediums are endoagar and nutrientagar. The petri dish and nutritional medium are sterilized before use. Next the sample that you are investigating is introduced to the dish. Some nutritional substrates are formulated such that only predetermined organisms can grow on them and others not. If that happens, you know at once that those organisms are present in the sample.

OUTBREAK

An outbreak is the appearance of a disease to a greater extent than expected at a specific place and given moment. The word implies nothing further with regard to the scale. Two related cases of a rare infection can already be enough to form an outbreak. Thus a well-localized small group of patients can just as easily be involved as a thousand people spread over a country or world region. Outbreaks that strike an entire country or region are epidemics. Outbreaks on an international and intercontinental scale are called pandemics.

The American CDC has described a series of steps you can follow if you are investigating an outbreak:
· Check the diagnosis of the disease involved.
· Check if there is truly an outbreak, and if the group of sick people is normal for the location and the season.
· Determine who will be regarded to be a case.

- Map the spread of the disease using information technology.
- Formulate a hypothesis for the cause of the outbreak.
- Check the hypothesis by means of collected data and the analysis thereof.
- Adjust the hypothesis and continue to check it.
- Develop an intervention and prevention strategy and apply it.
- Share your findings with the community.
- After it is over, organize a debriefing.

Diverse outbreak patterns exist that can be useful to identify the way in which an outbreak will spread and/or its source, and to predict the future number of cases. Thus there are COMMON SOURCE OUTBREAKS, in which the victims fall ill from the same source (for example a communal pump that has contaminated water), and PROPAGATED OUTBREAKS, in which there is transmission from person to person. Outbreaks can be linked to RISKY BEHAVIOR (as in the case of a sexually transmitted disease) or be zoonotic (caused by a pathogen which is usually present in animals).

PANDEMIC
See: AIDS, outbreak

PATHOGEN
Disease-causing agent. PATHOGENIC: causing disease

PATIENT ZERO
See: index case

PCR
See: polymerase chain reaction

PLASMA
See: blood

PLASMAPHERESIS
Plasmapheresis is a technique in which blood plasma is drawn from a patient but the donor retains the blood cells or gets them back. Plasmapheresis can be used in different situations. For example, the blood of a patient might contain antibodies that are inflicting damage and can be removed with the plasma. In this case, the patient receives donor plasma to replace that removed. Sometimes healthy people undergo plasmapheresis so that their plasma can be used for the production of blood products. We applied the technique in 1976 to obtain antibodies from cured Ebola patients, with which new patients were then treated.
See: blood

PLASMODIUM FALCIPARUM
See: malaria

POLIO

In full: poliomyelitis or infantile paralysis. Polio is caused by the polio virus that is shed when infected people defecate. With insanitary conditions and poor personal hygiene, the virus can then be transmitted from person to person. Most infections aren't even noticed, but somewhat less than 1 percent of infected people suffer muscular paralysis, which can be untreatable and lead to malformations in growing children. People also die from polio due to breathing problems.

There are two polio vaccines, the dead Salk vaccine and the living Sabine vaccine. The Salk vaccine is safer but was previously also more expensive and troublesome, because it must be injected by a doctor or nurse. Now that the Salk vaccine has been incorporated in the DWTP injection (against diphtheria, whooping cough, tetanus, and polio), these problems have largely been resolved.

The Sabine vaccine contains a weakened but living polio virus. It is usually taken by mouth, on a lump of sugar for example. The disadvantage is that people with a bad defense can become seriously ill from it. Until 2001 Belgium used the Sabine vaccine but then switched to the Salk. The Netherlands already used the Salk vaccine before the DWTP injection.

In 1988 the World Health Organization began a campaign to eradicate polio. Since then the number of new cases has plunged spectacularly: from 350,000 cases at the beginning to 406 cases in 2013. The disease is now only present in Pakistan, Afghanistan, and Nigeria, due to religious scruples and violence directed at vaccination teams. In Syria the vaccination program has been thwarted by the war and there have again been new cases since 2013.

In Belgium polio vaccination has been recommended since 1958, and legally mandatory since 1967. Due to this, no new polio cases have been reported since 1970. The last great polio epidemic in the Netherlands occurred in 1956, when almost 1,800 people became paralyzed and 70 patients died. After this there was a mass vaccination campaign. In 1971 there was an outbreak of polio on the Veluwe and in Staphorst. As at present, the last polio outbreak in the Netherlands took place in 1992 in Streefkerk. These two outbreaks struck so-called "conscientious Calvinists," people who do not allow their children to be vaccinated out of religious conviction. Treatment of disease is allowed. Otherwise the Netherlands has a high level of vaccination and there is group immunity, by which non-vaccinated people are also protected, seeing as the virus cannot spread itself through a population in which the vast majority is immune to it. Group immunity does not exist in the "Bible belt," an area that runs from Zeeland to the west of Overijssel. Indeed, in 1999 and 2013, there were also measles epidemics in this zone.

POLYMERASE CHAIN REACTION (PCR)

A technique to multiply specific parts of extremely small quantities of genetic material so that there is enough to investigate the DNA and RNA. With PCR you always need a starting point. You must define what you wish to multiply and investigate, and you must know how the DNA sequence that you are seeking starts and ends. These pieces are called "primers."

PCR is used in forensic investigation, for example, to identify someone by means of the DNA in a speck of blood. The technique can also be applied to check if there are genetically modified organisms in foodstuffs. And in many infectious diseases you can use PCR to reveal the responsible pathogen. PCR is usually not only more sensitive and specific (and thus more reliable), but also quicker than other methods.

See: Ebola tests, virus test

RESISTANCE

Resistance against an antibiotic means that a bacterium can no longer, or at least less effectively, be neutralized by it. This can happen because the antibiotic doesn't work well or is badly used, such as being administered over too short a time. In both cases the chance exists that strong bacteria survive the treatment, multiply once again, and transmit their resistance to the antibiotic through their genetic code to their offspring. If that happens sufficiently often, natural selection sees to it that all such bacteria resist a specific antibiotic and perhaps also similar antibiotics.

Resistance is a worldwide problem. An example is *Staphylococcus aureus*. It already became known in 1947—after four years of mass production of penicillin—that this bacterium had developed resistance to that antibiotic. Medicine switched to methicillin, but in 1961 the first Staphylococcus aureus bacteria were found that had also become resistant to methicillin. They received the name MRSA (methicillin-resistant Staphylococcus aureus). Staphylococcus aureus is present in 30 percent of people on the skin and the mucosal membranes. The bacterium is usually innocuous, but it can cause infections in people who have a weakened defense. You can fight these infections with antibiotics, but in the case of MRSA, that no longer works. MRSA has become a notorious cause of hospital infections.

The most important causes of resistance are:
- The large scale use of antibiotics, including for cold and flu, ailments for which antibiotics are useless. Doctors often prescribe antibiotics because they aren't sure of the cause of an illness, because they wish to prevent a more dangerous bacterial infection following a banal viral infection, or because the patient insists strongly on an antibiotic and if need be will go to another doctor to get it.
- An incomplete course of an antibiotic. A patient recovers quickly due to a prescribed antibiotic and stops taking it too early, against the doctor's prescription, at a time when not all the bacteria have been killed.
- Long courses of antibiotics in low doses.
- The systematic preventive use of antibiotics in livestock production.
- Great mobility, whereby resistant bacteria are spread worldwide.

Another example is the ESBL bacterium. ESBL stands for Extended Spectrum Beta-Lactamase, an enzyme that can break down certain antibiotics. The ESBL enzymes are produced by bacteria in the intestines and in this way these bacteria are resistant to cephalosporin and penicillin, the antibiotics that the enzyme can break down. The bacteria *Klebsiella* and *Escherichia coli* can produce ESBL and thereby cause serious infections in people with a compromised immune system.

Viruses can develop resistance much more easily, not against antibiotics because they are already invulnerable to these drugs, but against anti-viral medicines. Resistance was a great problem at the start of the fight against AIDS. The answer was to combine three different medicines, such that the one drug attacked the viruses that another or both of the other drugs could not neutralize.

RESTV

Reston-Ebola virus, a species of the Ebola virus that appears not to be dangerous for humans.

See: Ebola virus

RNA

Ribonucleic acid (sometimes ribose nucleic acid).

See: Ebola medicines, Ebola tests, Ebola virus, hantavirus, virus, virus test

SERENDIPITY

The phenomenon in which you discover something important or useful that you were not seeking, or—as the American scientist Julius Comroe once described it—you are searching for a needle in a haystack and find a lusty farmgirl.

Serendipity is more important in science than many people realize. Countless discoveries have been serendipitous. This doesn't make such finds less admirable, because serendipity "befalls" wide awake, open minds that do not simply dismiss happenstance as happenstance and expect the unexpected. Thus serendipity should not be confused with a lucky shot, although that often happens. Serendipity is to see bridges where others see gaps, to recognize a pattern in seeming disorder, to see creative relationships between things where others register mere chaos. Happenstance is an event, serendipity a skill.

A famous example is what Alexander Fleming did, when during the cleaning of his laboratory in 1928 he noted that there were far fewer staphylococcus bacteria growing around colonies of mold on old culture plates. He ferreted away at it and discovered that the mold released a substance that hindered the growth of the bacteria: penicillin! To be sure it must have often happened that such a dish was lying about in one lab or another, but Fleming was the first to pay attention to it.

Another funny case of serendipity is the discovery of the sweetener aspartame by a researcher who was experimenting with a possible treatment for stomach acid: he licked his finger to turn the page of a textbook, and tasted something remarkably sweet.

Also America, X-rays, champagne, sticky notes, teabags, Viagra, the magnetron, and Teflon were discovered by serendipity.

Although most fruits are plucked by science as a result of purposeful research, it is a characteristic of the true scientist to assume that mere coincidence is the cause of every visible relationship until this has been eliminated, and yet to suspect a pattern of cause and effect behind every obvious coincidence until the opposite has been proven. Serendipity is an important but underestimated supplement to traditional scientific thought and method, which is based on logic and predictability. The art is to observe, to evaluate, and to repeat in a focused, disciplined, and purposeful manner, but at the same time to remain open to the possible chance of the unexpected. Serendipity is a prophylactic against narrowing of the mind and the complacency of established science. And although you cannot utilize serendipity on purpose—you are, by definition, searching for something that the phenomenon won't deliver—the chance of serendipity can be heightened by honing one's attitude during discussions of scientific research, until it becomes second nature. It can also help to share scientific data and procedures with other people who possibly look at them differently, and to regularly wander down untrodden paths that have formerly been regarded to be dead ends. Serendipity has not only delivered much benefit to scientists, but also to inventors, industrialists, and artists.

The word comes from Serendip, an old name for Sri Lanka or Ceylon, where, according to a fairytale, princes lived who discovered things for which they were not searching.

SERUM/SERA

Serum is blood plasma without clotting factors. Sometimes serum (plural: sera) is used when in fact "antiserum" is intended: serum in which there are many antibodies against a poison or pathogen. Thus antisera are used against snakebites. Antisera from people who have survived a bacterial or viral infection and thus have antibodies against the infection can be used to treat new patients. However, this can provoke a serious allergic reaction. Therefore, if it is at all possible, pure antibodies are now used instead of (anti)sera, preferably monoclonal antibodies that are made in the laboratory.

In 1976 we treated patients with the sera of other Ebola patients who had survived the disease, such as Sukato. Sometimes this worked, sometimes not. Before we knew for certain that we were not dealing with a species of Marburg but a wholly new virus, and before we had sera from cured Ebola patients, we also tried to treat an Ebola patient with the serum from a recovered Marburg patient because it was the only possibility, but that didn't work.

See: blood, plasmapheresis

SEROPOSITIVE

See: AIDS

SMALLPOX

Smallpox (also called variola; note that the "big pox" was syphilis) was a highly contagious, life-threatening disease. I say *was* because it has been eradicated by science. Its last victim fell in 1978, a British woman who worked for the University of Birmingham, where research on polio was underway. She did not survive.

Since 1980 our planet has officially been free of smallpox. That could be achieved because there was a good vaccine against smallpox and there is no reservoir of the virus in animals. Thus it cannot hide anywhere. The variola virus is now only preserved in two laboratories for scientific purposes.

The cause of smallpox is the smallpox virus. Animals do not get variola but are vulnerable to viruses that resemble it, and which can even make people immune to human poxes. Indeed, this is how the principle of vaccination was discovered.

The virus raged for thousands of years and killed people or maimed them for life. European explorers exported the virus to exotic peoples that had never been in contact with variola and who had not the slightest resistance to it. They died from the disease on a massive scale, in some cases up to 90 percent of the population. This is also the most important reason why there are so many whites and blacks in North America, and so few Native Americans ("Indians"). In the eighteenth century smallpox was even deliberately used as a biological weapon to fight Native Americans. In the Cold War both the US and the USSR investigated the use of smallpox as a bioweapon. Certainly after 1980, with immunity diminishing in the wake of the disease being eradicated, this was an interesting avenue of research for biological warfare.

Two species of the smallpox virus exist: the relatively mild form Variola minor and the dangerous Variola major, which, sadly, was much more prevalent, and which in turn also had a number of variants.

The most important symptoms of smallpox were firstly a high fever and aching muscles—during this phase a patient was highly infectious—and after about a week little blisters appeared on the skin, first on the forehead and within a day or two over the whole body. Thus the little blisters were not limited to the arms and legs, as is the case for the much more innocuous chickenpox or varicella, which is caused by another virus (the varicella-zoster virus). In the blistering phase the patient felt somewhat better. The blisters were not filled with pus but with destroyed skin tissue. In a third phase the patient again became sicker, the blisters emptied and became scabs that later left scars, and then the surviving patient was no longer infectious and was immune. Ten of every hundred people with variola major died from the disease. Sometimes there were so many blisters that they coalesced into large areas that destroyed the skin like burn wounds. Sixty percent of such patients died.

There was also the so-called "flat type variola major," with blisters that contained less water, struck children in particular, and was almost always fatal.

A rare, hemorrhagic form of smallpox also existed, with bleeding beneath the skin, which thereby turned black. Black smallpox had a CFR of almost 100 percent.

Variola minor caused a rare, milder form of the disease with a lower fever and a CFR of less than 1 percent.

While there was no treatment for smallpox, there was indeed a vaccine, and with it the disease was eliminated. My good friend Stan Foster contributed significantly to this, and I, in my own manner, also contributed a little. During my expeditions to possible reservoir areas of Ebola I also looked for indications of smallpox and monkeypox.

STRAIN

A strain of a micro-organism or virus is a collection of entities that are descended from an individual micro-organism or virus. Entities of the same strain thus have the same genetic characteristics as the original micro-organism or virus. Within one species of the Ebola virus, EBOV, for example, you have countless different strains of the same species. By investigating the genetic material of the virus in a sample, you can check which strain or strains you are dealing with. On the basis of such research it could be determined, for example, that the epidemic in West Africa was unconnected with the simultaneous outbreak in the Congo. In sample after sample the cause was EBOV, but different strains.

See: Ebola medicines

SUDV

Sudan virus, a species of the Ebola virus.

See: Ebola virus

TAFV
Taï Forest virus, a rare species of the Ebola virus.
    See: Ebola virus

T-HELPER CELL
See: defense

TUBERCULOSIS
Tuberculosis or TB is a contagious disease that is usually caused by *Mycobacterium tuberculosis*, a bacterium that was discovered in 1882 by Robert Koch. Thus the bacterium is also called the Koch bacillus. The name tuberculosis comes from the tubercles, or little bumps, that are typical during inflammation caused by these bacteria.

Earlier, near the beginning of the previous century, almost everyone was contaminated with Mycobacterium, but most people did not know, healed, and were from then on immune, although sometimes concentrations of the TB bacteria remained at certain sites in the body, and these could cause problems when the body's defense was severely burdened. People who became sick following an infection often had to go to a sanatorium by the sea or in forests. Nevertheless it often ended badly. TB was a bit like AIDS at the end of the previous century: a disease that was difficult to treat and which was frequently fatal.

Tuberculosis affects the lungs especially (pulmonary tuberculosis), but can in principle attack all organs. The most important symptoms of pulmonary tuberculosis are coughing (sometimes with blood), weight loss (hence "consumption"), and pain in the ribcage. According to the World Health Organization, in 2013 approximately nine million people had TB, of whom 1.5 million died. Roughly a quarter of those killed by TB were HIV positive. TB worsens HIV infection and HIV infection helps TB to proliferate in the body.

Worldwide, TB is gradually diminishing. It has been estimated that 37 million human lives were saved between 2000 and 2013 by effective diagnosis and treatment, but precisely because the disease can be treated, the 1.5 million deaths recorded in 2013 is still a high figure. In Belgium there are about one thousand new cases per year. Approximately half of these patients are not Belgian.

Infection usually occurs through breathing in the bacteria that are floating in the air and which are spread by pulmonary tuberculosis patients when they cough. Previously you could also get TB from contaminated cow's milk. Although resistant TB is becoming an ever greater problem, the disease can usually still be treated successfully with antibiotics. The catch is that treatment takes half a year and can have side effects.

VACCINATION
The word "vaccination" comes from the Latin *vacca*, which means "cow." In the eighteenth century the British doctor Edward Jenner noted that dairymaids who had been infected with cowpox during milking were never infected by smallpox. And Turkish farmers had discovered that you never develop dangerous smallpox if you allow yourself to be infected by someone with a mild form, such that you get this form of the disease yourself and are protected against the serious type.

At birth you receive antibodies from your mother, but these disappear after a couple of months. From then on a baby must produce antibodies itself, and thus first be infected and possibly fall ill, with all the risks that are involved. In a vaccination non-damaging disease-causing agents are introduced to the body, so that man or beast makes antibodies without first having to become sick.

Different types of vaccines exist: vaccines with living but weakened organisms (such as some vaccines against polio and abdominal typhus), and vaccines with pieces of dead micro-organisms. The pieces could come from a real pathogen (as in the vaccine against diphtheria) or from a laboratory that makes synthetic vaccines (as in the hepatitis B vaccine).

Especially in earlier times vaccines were made with dead or weakened viruses. The danger was that people would still fall ill because the "dead" virus might still contain a few living viruses, or because the attenuated virus regained its strength. Moreover not all viruses could be cultured in sufficiently large quantities for the production of vaccines.

Today the medical world makes use of the so-called "recombinant vaccines." This type of vaccine consists of the combination of a vector and pieces of genetic material from the virus. The vector is a piece of a virus other than that against which the vaccine is targeted. Thus parts of a certain adenovirus are often used as the vector. These are safe and can be made relatively easily. An extra advantage of adenovirus vectors is that they do not only stimulate the production of antibodies but also a cellular immunological response. (For cellular defense, see: defense.)

Most vaccines are injected, but some live vaccines must be swallowed or introduced to the body via a little cut in the skin. Often one or two repetitions (boosters) are required.

Some vaccines can be given to everyone. In Flanders only the vaccine against polio (infantile paralysis) is mandatory, but within the so-called "basic vaccination scheme," people are currently routinely vaccinated against mumps, diphtheria (croup), Haemophilus influenzae type B (a bacterium that, among other things, can cause meningitis), hepatitis B, pertussis (whooping cough), measles, meningococcus type C, pneumococcus, polio, rubella (German measles), tetanus, and the human papilloma virus. The last-named can cause cervical cancer (cancer of the cervix). The HPV portion of the vaccination scheme is only for girls. In the first year of junior high school, girls can have themselves vaccinated for free.

There are other vaccinations for specific risk groups. Thus people in health care, the aged, diabetics, and lung patients benefit from flu jabs.

People who are traveling to poor or tropical countries can protect themselves with vaccinations, particularly against diseases such as diphtheria, tetanus, and polio with the DTP vaccine, or against hepatitis A, yellow fever, and abdominal typhus.

Although obstinate global anti-vaccination lobby groups exist which link vaccination with needless side effects, allergic reactions, and disorders such as autism, it appears from studies that vaccines generally have few side effects and that the advantages outweigh the side effects. There are other people who are opposed to vaccination on account of their beliefs or their outlook on life. However, it is a fact that few medical practices have saved as many lives as vaccination. The fact that unvaccinated people often do not fall ill is indeed partly thanks to the fact that other people around them have had themselves vaccinated.

See: defense, immunization

VERO CELLS

Vero cells are cell lines (collections of identical cells) which are used in cell cultures. They were isolated by Japanese researchers in 1962 from the kidneys of an African monkey and since then have been produced on a large scale for laboratories everywhere in the world. You can still order them today. Among other things, vero cells are used to culture viruses, for example to see if there is indeed virus in a certain sample or to see if such a virus reacts with an antiserum.

The special thing about vero cells is that you can allow them to multiply time and again, without them changing through age. Moreover they do not produce any interferon, in contrast with normal animal cells, when they are infected with a virus. That is equally good, because interferon can kill viruses and that is unfavorable if you are trying to culture viruses or test the efficacy of a certain substance against the viruses.

VIRUS

A virus is an extremely small amount of genetic material encapsulated by proteins. Just like a bacterium, a virus can penetrate plants, animals, or people and make them sick, but beyond this viruses differ greatly from bacteria. An important difference is that viruses are much smaller. The diameter of most viruses lies between 20 and 300 nanometers (one millionth of a millimeter). Thus you need an electron microscope to see them. Indeed, not long ago, giant viruses of 750 nanometers were discovered, but that is still hundreds of times smaller than bacteria, which are usually 1 to 5 micrometers in diameter (a micrometer is one thousandth of a millimeter). Yet another important difference with bacteria is that antibiotics do not work against viral infections. For many viral infections there is not even a single therapeutic treatment. Moreover—a third big difference—bacteria are living, viruses not really.

VIRUSES: LIFE AND DEATH

Actually, you can't call a virus an "organism," although in practice that often happens. In biology an organism is an entity that displays the characteristics of life, and viruses do not have all such characteristics. An individual virus does not grow, does not exchange substances with the environment, and does not produce waste products. They cannot reproduce themselves as individuals or together with other viruses.

If viruses do not "live," you ought to be able to describe them as "dead matter." Yet that is also not quite right, because in contrast with a pebble or an ice crystal, a virus can adapt itself and in this manner maintain and improve itself. Thus there is definitely a kind of evolution in viruses. This evolution progresses even more quickly in viruses than it does in living organisms. Living beings are programmed to protect themselves and to multiply. That isn't the case for dead material, but for viruses the situation is more complicated. True enough, they are not really living, but they have found a way to propagate themselves and to spread. Thus viruses lie somewhere between living and dead matter.

How can a virus "adapt" itself? Well, the answer is that when it enters the living environment of an organism, a huge amount happens between that organism and the virus. The result of the exchange is that with time there are more viruses inside the organism, and among the new examples there are changes (mutations). Thus, although they do not reproduce independently, viruses do indeed change and their numbers increase. To do this they require a host. That is not necessarily always disadvantageous for the plant or animal involved. Sometimes the infected organism can live relatively normally and the virus remains dormant within it, or multiplies without causing too much damage. In other cases the virus makes the organism ill or kills it.

A whole lot happens without us noticing. It depends on the infected organism whether a disease-causing or deadly virus is recognized. It is a reasonable gamble that tens of thousands of species of virus exist that we have not yet discovered or will never discover, because we can see no sign of them. Only if the sick or dead organism is a person, a known and visible animal, or a valuable plant (i.e. a decorative or food plant) do the effects of the virus become conspicuous enough and it can be discovered or identified.

INTERACTION BETWEEN VIRUS AND HOST

What happens when a virus comes into contact with a host organism? Then it attaches itself to a cell of that organism and it injects its own genetic material into the cell, or fuses with it, so that it can use the cell as a virus factory. Viruses come into contact with all sorts of cells, but often this process only occurs when they contact specific cells. There must be a match between the protein capsule of the virus and the little parts on the outside of the cell that play a role in defense. If there is a match, the virus injects its own genetic material into the cell, the cell is disrupted, and the defense system no longer works so well. The cell then breaks apart (lysis) or dies (apoptosis). In some cases it goes differently and a virus injects useful material into the cells of certain organisms, but often viruses are bad news.

CONSTRUCTION

How is a virus constructed? Inside there are nucleic acids, usually those named DNA or RNA. This is the hereditable material of the virus. This genetic material is kept together by a capsule of proteins, also called a capsid. The capsule has two functions: it protects the virus from destruction by antibodies and it enables the genetic material to enter a host cell. Occasionally there is an extra layer around the capsule, called the envelope, but this is only present in some animal viruses.

CLASSIFICATION

Currently, more than 2,800 species of viruses are known. These are grouped in 455 genera, 103 families, and 7 orders, although tens of virus families have yet to be assigned to a specific order. A few examples of virus families are:
· adenoviridae (which usually cause mild problems such as airway infections and inflammation of connective tissues)
· filoviridae (among which are the Ebola and Marburg viruses)

- herpesviridae (the cause of chickenpox, shingles, and the various forms of herpes, among others)
- noroviridae (important causes of diarrhea)
- orthomyxoviridae (among others the influenza or "flu" virus)
- paramyxoviridae (measles, mumps, flu, avian flu...)
- papovaviridae (such as the papilloma virus that can cause cancer of the cervix)
- picornaviruses (such as the rhinovirus that causes the common cold, but polio and hepatitis A are also the result of picornaviruses)
- reoviridae (including the rotavirus which also causes diarrhea)
- retroviridae (HIV, for example, the cause of AIDS)
- rhabdoviridae (among others the rabies virus)

VIRUS TEST
There are different ways to check that you are dealing with a virus and to know which virus is involved.

MICROSCOPY
Under a microscope you can see what is happening in infected cells. To see the virus itself, you need an electron microscope.

VIRUS CULTURES
You can introduce a tissue or fluid sample into cell cultures in which it is possible that the virus may be able to multiply. This can help to determine which virus has caused the infection, because the cultured virus brings about changes in the cells that you can observe under a microscope.
    See: vero cells

MOLECULAR TECHNIQUES
With these techniques it is possible to identify the genetic material of a virus (DNA or RNA) in the feces, a piece of tissue, or the blood of someone with an infection. The composition of that code is always unique to each virus. Molecular techniques are important when culturing a virus would take too long or there are no other reliable methods to identify the virus. Two types of molecular techniques exist: hybridization and amplification.
    In hybridization, one checks to see whether a DNA or RNA string from a known virus (the probe) binds with a known piece of the DNA or RNA string of the virus that one wishes to detect (the target). If this happens, the combined hybrid can be detected by means of a so-called reporter molecule, which is part of the probe. The reporter molecule can be radioactive and emit particles, emit light by fluorescence, or act as a dye. Thus if unattached probes can be removed from a mixture of probe and target molecules, one can see if any probes have been bound to the target DNA or RNA, revealing the identity of the target.
    The most used amplification technique is the polymerase chain reaction or PCR, a sort of molecular photocopier. It makes it possible to multiply even minuscule amounts of a virus in a specimen and thus to detect it. To detect RNA viruses, a separate tech-

nique is required: RT-PCR (reverse transcriptase polymerase chain reaction). Other amplification techniques are NASBA and the so-called branched probes technique.

In serological investigations, antibodies are sought in the blood against antigens alien to the body, which come from a virus. These antigens switch the defense system on so that it makes antibodies which are directed against that specific micro-organism, and no others. Usually only two types of antibodies are measured: IgM and IgG. IgM antibodies are made first and thereafter IgG. An IgM antibody is only made if there was contact with an antigen a short time ago. By contrast IgG antibodies can be found in the body for years. Thus a positive IgG test can occur not only if there is an active infection that has already lasted a long time, but also if an infection has been cured. The concentration of the antibody in the blood is called the "antibody titer." Serological tests can be used to check if someone has contracted a specific virus or is immune to it.

Most serological investigations occur in two stages. First, you let the unknown antibodies in the blood specimen react with a sample of the virus that you suspect to be present. You subsequently study the interaction, possibly after you have added a dye or an enzyme to make the interaction visible. The interaction could cause fluorescence (a form of light emission), coloring, or agglutination (bunching together).

ELISA (enzyme-linked immunosorbent assay) is a technique that is often used. ELISA is a test in which an antigen or antibody is coupled to an enzyme to check if there is a match.

Other techniques are the complement binding reaction, indirect immunofluorescence, the hemagglutinin test, and the neutralization test.

See: Ebola tests

WORLD HEALTH ORGANIZATION (WHO)

The World Health Organization or WHO is a specialized agency of the United Nations, with its headquarters in the Swiss city of Geneva. The purpose of the WHO is to lay health and health care issues throughout the world on the table, to coordinate health care, and to improve health everywhere in the world.

Since the WHO was founded shortly after World War II the agency played an important role in the eradication of smallpox. Right now the priorities are contagious diseases (especially AIDS, Ebola, malaria, and tuberculosis), to limit the results of non-contagious diseases (such as heart and vascular conditions and diabetes), sexual and reproductive health, aging, food security, occupational diseases, and drug use.

The WHO issues the World Health Report, an important international publication about health, and the global World Health Survey. World Health Day falls on April 7 each year, the date on which the organization was founded in 1948. The WHO uses this opportunity to focus attention on a specific health theme. The organization also controls the classification of medical conditions (ICD-10) and means of treatment (ATC-coding).

There are 194 countries that have underwritten the goals of the WHO and are member states of the organization. The WHO has 8,500 people serving in 147 lands. The WHO received much criticism during the Ebola epidemic of 2013–2015, from Doctors Without Borders, among others, because the organization responded too late and too feebly to the crisis. The WHO has itself conceded this.

YELLOW FEVER

Yellow fever is an infectious disease that strikes monkeys, apes, and people, and which is especially prevalent in Sub-Saharan Africa, with the exception of Southern Africa. Yellow fever is also present in Central America and in the north of South America (5 percent of cases).

The sickness is caused by the yellow fever virus, which is transmitted by certain mosquitoes—in Africa especially by *Aedes aegypti*, the species of biting mosquito that can also transmit the dengue and Chikungunya viruses among others, and by *Aedes africanus*.

Only a quarter of people fall ill after infection. Symptoms begin after three to six days and usually include a high fever, headache, aching muscles, and nausea. If people do not recover from this, they enter the second, toxic phase, of which the most important symptoms are recurrent fever, jaundice as a result of liver damage (hence the name "yellow fever"), and bleeding (from the mouth, eyes, and in the digestive system). Due to this, Ebola and Marburg can easily be confused with yellow fever. Thus the original diagnosis during the Yambuku epidemic of 1976 was yellow fever.

About 15 percent of all infected people enter the advanced phase of the disease and one in five of these patients die, which brings the overall CFR of yellow fever to 3 percent. Roughly two hundred thousand people contract yellow fever annually, of who approximately six thousand do not survive. The numbers have increased over the past two decades. There is no therapeutic treatment for the disease, only treatment of the symptoms. However, there is a vaccine. In many countries where yellow fever is present, it is compulsory for visitors to be vaccinated.

ZMAPP

See: Ebola medicines

ZOONOSIS

An infectious disease that can be transmitted from animal to human. That can happen via bacteria and viruses, but also via parasites, prions, fungi and molds, worms, arthropods, etc. While the course of the disease in people is often severe, the disease-causing agents (pathogens) do not cause too much damage in their normal hosts—otherwise the hosts would quickly die and the pathogens would lose their shelter—but in intermediate animals and people, the pathogens can provoke a sharp defensive reaction.

Zoonotic pathogens reside in vertebrate animals and infect people intermittently. The animals are their reservoirs; this is where they hide. A pathogen can be transmitted directly from its reservoir to people, or via other animals. In western countries zoonoses are often transmitted via pets or livestock, because these are the animals with which we most often come into contact. Zoonoses can thus be directly transmitted to people from other vertebrates, but often there is no contact between animals and people and the transmission occurs via an animal product (a dairy product or meat, for example) or via insects. Insects are highly mobile and this allows them to transmit diseases even if the vertebrate host and humans are not near each other. The last organism to carry the pathogen before it is transmitted to people, or the actual means of transmission, is called the vector. The actual means of transmission could be direct contact, an insect bite or a sting, for example.

Ebola, among others, is a viral zoonosis. Often Ebola outbreaks begin when people kill a sick monkey or ape, or find a dead one, and subsequently butcher or touch it in some other way. Because certain monkeys and apes themselves fall ill and often die from Ebola, they are just a stepping stone and the true reservoir consists of other animals. It is possible that the virus survives in fruit bats. Other viral zoonoses are Lassa fever (transmitted by rodents), foot-and-mouth disease (transmitted by cattle, sheep, goats, pigs, and other even-toed ungulates), and rabies (transmitted by dogs, foxes, bats…). Among other bacterial zoonoses are Malta fever (brucellosis), anthrax, and the plague, with rodents being the reservoir and fleas (or flea bites) as the vectors. An example of a zoonosis caused by a prion is Creutzfeldt-Jacob disease, with cattle being the reservoir. The vector is the eating of cattle brains or nerves. Ringworm is a fungal zoonosis. Thus the disease has nothing to do with worms but with a fungus that occurs in humans and that you pick up through contact with dogs or cats. Then again, toxoplasmosis is an example of a zoonosis that is caused by a parasite. You can ingest the parasite via cat feces (through contact with the cat litter, for example), but also by eating raw or undercooked meat or unwashed vegetables, which have been touched by the droppings of infected animals.

The reverse of a zoonosis also exists, anthroponosis, which is a disease that is transmitted from people to animals. Of course this happens less than vice versa, because we do not bite and sting animals. Transmission is usually by skin contact. An example of an anthroponosis is chytridiomycosis, a fungal disease that people can give to frogs if they touch them with their bare hands. Many human diseases can be transmitted to apes and monkeys because they are so similar to us, not only in appearance but biologically, too. Zoos know this and take precautionary measures. Hence apes are sometimes given the same vaccines as humans. There have already been diverse Ebola outbreaks among gorillas, in which up to 90 percent of the animals in a specific region have died, but so far as is known Ebola has never yet been transmitted from a person to a gorilla.

To conclude, amphixenoses strike both humans and animals and are transmitted in both directions.

SOURCES

PRINTED

- *Ebola: The Natural and Human History of a Deadly Virus* (David Quammen, 2014)
- "Ebola Virus Hemorrhagic Fever," Proceedings of an International Colloquium on Ebola Virus Infection and Other Hemorrhagic Fevers, held in Antwerp, Belgium, December, 6–8, 1977 (S.R. Pattyn, editor, 1977)
- *Ebola: What To Do?* (Sudhir Bhatia, 2014)
- "Een brug tussen twee werelden: Het Prins Leopold Instituut voor Tropische Geneeskunde Antwerpen 100 jaar" ["A bridge between two worlds: 100 Years of the Prince Leopold Institute for Tropical Medicine Antwerp"] (ITM, 2006)
- *Epidemics: Deadly Diseases Throughout History—Ebola* (Allison Stark Draper, 2002)
- *Filoviruses: A Compendium of 40 Years of Epidemiological, Clinical, and Laboratory Studies* (J.H. Kuhn and C.H. Calisher, 2007)
- *Geen tijd te verliezen* [*No Time To Lose*] (P. Piot, 2012)
- *The Coming Plague: Newly Emerging Diseases In A World Out Of Balance* (Laurie Garrett, 1994)

ONLINE

- "2014 Ebola virus disease epidemic timeline," Wikipedia
- "Alles over Ebola" ["All about Ebola"], *Knack*
- *deredactie.be* from VRT
- "Ebola," *De Standaard*
- "Ebola blog," Health Home Capacity
- "Ebola Outbreak Full Coverage," ABC News
- "Ebola Outbreak in West Africa," European Centre for Disease Prevention and Control
- "Ebola Research Center," *The Lancet*
- "Ebola Update," Government of Liberia
- "Ebola virus disease," Wikipedia
- "Ebola virus epidemic in West Africa," Wikipedia
- "Eboladossier: Ebola-epidemie in West-Africa," Artsen Zonder Grenzen ["Ebola Dossier: Ebola Epidemic in West Africa," Doctors Without Borders]
- "Fact Sheet: African Union Response to the Ebola Epidemic in West Africa," African Union
- "Global Alert and Response," World Health Organization
- International Health Policies, blog of the ITM
- Johns Hopkins Ebola Forum October 14, 2014
- "La découverte du virus ebola" ["The Discovery of the Ebola Virus"], Pierre Sureau, 1976
- Medecins Sans Frontières International [Doctors Without Borders International]
- *The Guardian*
- Twitter accounts: Ebola Alert, WHO, CDC
- Virology Professionals LinkedIn discussion group

ACKNOWLEDGMENTS

I thank my wife, Constantia "Dina" Petridis—my harbor, the mother of Ilse and Sonja—for her determination to preserve the unity in our diversity. For this, too, she earns a high distinction, and thus not only for her studies in chemistry.

I thank my children, their partners, and our grandchildren, Lore and Jana, for their affection and support, despite the fact that I was seldom present during my daughters' childhoods. It was Ilse who came up with the idea of making a book out of the breast pocket cards that I had filled during my many travels.

I thank my ex-secretary Ciska Maeckelbergh, who on top of her work made time to type out my pocket cards to create a 460-page document.

I thank Christine Caals, a friend of Ilse and one of her colleagues at the Flemish AIDS coordination center IPAC. My daughter knew that Christine writes fluently, and so after a couple of telephone calls and e-mails in September 2014 we drove to see Christine, who lives in the Ardèche. On the way we stayed in Cahors, where Yvette Baeten, another former secretary of mine, runs a fantastic B&B. Between Cahors and the Ardèche the name of medical writer and scientific journalist Marc Geenen came to mind. He had interviewed me at different times about HIV and Ebola, and could perhaps help to give the book a journalistic character and explain scientific concepts in common language, which I find very important. During our days in the Ardèche, Christine interviewed Dina and me about our lives. I had the film footage that I had shot in 1976 during the Ebola outbreak in Yambuku. Christine agreed at once to write a book. On an Indian summer Sunday in October, during a cracker of a conversation on my terrace in which Paul Geenen also participated (Marc's uncle), the structure of the book took shape. For many years I had worked together with Paul in the field of HIV diagnostics. He is a good friend, with a wide range of interests and a keen analytical ability. For that matter, he is one of the people who checked the proof copy, alongside Kevin Ariën. The English proof copy was read by Judy and John Salter-Best and Janet Tobias. To them, too, I am deeply indebted.

I thank Katrien Van Oost of my publisher Lannoo, who immediately showed interest in our idea. Katrien was the essential fourth wheel of the car for our writing triumvirate. Katrien also supported the idea to publish the English version of the book. In this I was, above all, the inkpot into which my inspired co-authors dipped their pens. They did so excellently. I thank Christine, Marc, and Katrien with all my heart. Special thanks to Mark Swanepoel for the excellent translation. I also thank the staff at *deredactie.be*. Their almost daily, high-quality online reporting of the current Ebola epidemic was a shining light in the tsunami of worldwide coverage.

Many thanks also to Bruno Gryseels, Director of the Institute of Tropical Medicine, Antwerp, Belgium, for the generous sponsoring of the English version of the book.

My thanks also goes to Roland Scholtabers, head of the communication service of the ITM, and his colleagues Nico Van Aerde and Jean-Pierre Wenseleers, who created a website especially for this book with historical images concerned with Ebola.

I thank my parents, Willy van der Groen and Maria Vingerhoets, for the love that I have received from them, the freedom to follow my own life, and the support upon which my family could always count. I am equally thankful to Paula Janssens, Dina's mother, who was of inestimable value to the balanced development of our daughters. Thanks to Paula, now and then Dina could accompany me on one of my trips, a favor for which we also thank Frits Parmentier, Dina's head of service at the University of Antwerp.

I thank the people who have given me a push in the right direction. It is thanks to Clement De Bruyne that I could start my biochemistry research work. At the end of the 1960s and start of the 1970s, Dina and I lived in Ghent and completed our doctorates there. We think back on the joyful work atmosphere there with warmth and gratitude.

I am very thankful to my colleague Françoise Portaels that she departed for Zaïre, so that for a while I could take her place at the Institute for Tropical Medicine. That occurred at the discretion of Professor Stefaan Pattyn, my service head, and the ITM Director P.G. Janssens and secretary Albert Graré.

In 1974 Peter Piot became a colleague of mine as an assistant in the Microbiology service of Stefaan Pattyn. His arrival heralded a stream of ideas and initiatives. It was on his impulse that we embarked on the study of HIV and AIDS in Africa after the Ebola adventure. Peter gave me an enormous boost, for which I am thankful. After Peter left, Marie Laga, Françoise Portaels, Luc Kestens, and I expanded what he had started. That work is now being continued by my successors Guido Vanham and Kevin Ariën.

This book would be of lesser value without the epidemiological notes made by my highly regarded colleague Pierre Sureau of the Institut Pasteur in 1976, when he—just like Peter and I—was a member of the international commission that fought and investigated the Yambuku epidemic. In *Yambuku: Une nouvelle fièvre hémorrhagique Africaine—La découverte du virus ebola*, he describes in great detail the activities of the Zairean doctors and nurses, of the Belgian mission sisters and fathers, and of the international team. I dedicate my book to Pierre, a great man who has left us much too early.

In conclusion, I thank everyone with whom I have worked through all the years, home and abroad, for their support, dedication, fellowship, humor, generosity of spirit, exceptional talents, understanding, and warm humanity. I have mentioned many of them in this book, but I would have to write another one to be able to mention everyone who I should thank for something.

Guido

LIST OF PEOPLE MENTIONED IN THE TEXT